PHARMACOLOGY – RESEARCH, SAFETY TESTING AND REGULATION

THE PHARMACOLOGICAL GUIDE TO METFORMIN

Pharmacology – Research, Safety Testing and Regulation

Additional books and e-books in this series can be found on Nova's website under the Series tab.

PHARMACOLOGY – RESEARCH, SAFETY TESTING AND REGULATION

THE PHARMACOLOGICAL GUIDE TO METFORMIN

NATALIE LAJOIE
EDITOR

nova
Medicine & Health
New York

Library of Congress Cataloging-in-Publication Data

ISBN: 978-1-53616-634-7
Library of Congress Control Number:2019954056

Published by Nova Science Publishers, Inc. † New York

CONTENTS

Preface

Almost six decades after its discovery, metformin still remains the gold standard drug for the management of type 2 diabetes mellitus, owing to its euglycemic property and cost-effectiveness. It is the most commonly prescribed and utilized oral antidiabetic drug worldwide. However, no two persons with the same phenotypical characteristics, such as age, gender, body weight and others, tend to respond similarly to equivalent doses of metformin. As such, The Pharmacological Guide to Metformin first focuses on the various genetic polymorphisms that influence the therapeutic effects of metformin.

Physicochemical, pharmacological, and pharmacokinetic properties of metformin will also be reviewed.

The side effects of this drug can be dangerous. Metformin-induced lactic acidosis is a lethal disorder. The first significant deviation is gastrointestinal tract involvement. Common infection or the effect of metformin itself causes vomiting and diarrhea, with subsequent hypovolemia causing pre-renal acute kidney injury, which in turn raises metformin levels up to toxic values. Development of kidney failure is often unexpected and fast.

Following this, the authors explore the differences between healthy and cancer cells and how this may influence the metformin effect on these cells.

In conclusion, the molecular mechanisms underlying the anticancer effects of metformin in breast and ovarian cancers are summarized, and the potential anticancer applications of metformin-based combinatorial chemotherapy are discussed.

Chapter 1 - After almost six decades of its discovery, metformin still remains the gold standard drug for the management of type 2 diabetes mellitus owing to its euglycemic property and cost-effectiveness. It is the most commonly prescribed and utilized oral antidiabetic drug worldwide. However, no two persons with the same phenotypical characteristics, such as age, gender, body weight and others, tend to respond similarly to equivalent doses of metformin. Moreover, a higher proportion of patients fail to attain optimal glycemic control with metformin. This variability in glycemic response to metformin could be attributed to the differences in the genetic architecture of the individuals. Various pharmacokinetic and pharmacodynamic genetic markers determine the ultimate glycemic response to metformin. Metformin is a charged hydrophilic molecule and it relies heavily on the membrane transporters to traverse across the biological membranes. It is not metabolized inside the human body and is excreted unaltered through the liver and kidneys. Hence, the optimal way to assess the pharmacokinetic involvement is to study the genes coding for the metformin drug transporters, which are involved in the absorption, distribution and elimination of the drug. The transporters of metformin include both the influx (which help the metformin to enter the cells) and efflux (which push the metformin outside the cells) transporters located at various cellular sites, namely, the intestine, the liver and the kidneys. The influx transporters include the OCT1 (organic cation transporter 1), OCT2, OCT3 and PMAT (plasma membrane monoamine transporter); the efflux transporters include MATE1 (multidrug and toxin extrusion protein 1) and MATE2. All these transporters are secondary active transporters and belong to the solute carrier (SLC) superfamily of transporters; the secondary active transporters obtain energy from the oppositely directed proton gradient, hence, they are also known as the H^+/organic cation antiporters. The OCTs are polyfunctional transporters which aid in the influx of metformin into cells. The OCT1 is located predominantly at the sinusoidal surface of the liver cells and also at the basolateral membrane of enterocytes; whereas, the OCT2 is expressed basolaterally in the kidney tubular cells. The OCT3 and PMAT are present at the luminal (apical) side of the intestinal cells. The metformin efflux transporters MATE1 and MATE2 are located in the luminal (canalicular) side of the liver and kidney, respectively; the MATE1 transporters are additionally present in the luminal side of the kidney. The facilitative bidirectional glucose transporter GLUT2 located in the liver, kidney, pancreas and small intestine is also implicated as one of the major genetic determinants of metformin response in large genome-wide association studies. The genes coding for OCT1, OCT2, OCT3, PMAT, MATE1, MATE2 and GLUT2 are *SLC22A1*, *SLC22A2*, *SLC22A3*, *SLC29A4*, *SLC47A1*, *SLC47A2* and *SLC2A2,* respectively. Studies done to assess the pharmacodynamic aspect of metformin, to predict the therapeutic efficacy and safety and to gain insight into its mechanism of action have shown that there are various potential genetic factors responsible for the variable therapeutic response of the drug. The most commonly implicated among these is *ATM* gene polymorphisms (ATM stands for *ataxia telangiectasia*); *ATM* gene encodes serine-threonine kinases that are involved in cell-cycle regulation and in the phosphorylation of AMP-kinase (AMPK) which is a major target for metformin. Polymorphisms in *ATM* gene can alter the AMPK levels which in turn can affect the glycemic response to metformin. Though the field of pharmacogenetics of metformin is amassing more and more wealth of knowledge the actual translation of the same to the clinical settings is suboptimal; the reasons being the multigenetic involvement and lack of adequately powered and scientifically robust studies. In this chapter, the authors focus on the various genetic polymorphisms that influence the therapeutic effects of metformin.

Chapter 2 - Metformin hydrochloride is a widely used biguanide derivative antidiabetic agent for the treatment type 2 diabetes mellitus. Metformin does not improve insulin secretion and produce normally hypoglycemia. Metformin decreases blood glucose levels by increasing skeletal muscle uptake of glucose, decreasing gluconeogenesis and absorption of glucose. Metformin has been found to be effective in the treatment of especially prostate, colon and breast cancer. Also, it is reported that metformin treatment decreases the risk of developing type 2 diabetes mellitus in women with polycystic ovarian syndrome. Metformin has a cationic charge at physiological pH and has very limited passive diffusion across cell membranes. According to Biopharmaceutics Classification System, metformin hydrochloride is a class 3 drug with high solubility and low permeability properties. Its absorption is slow and incomplete, and thus has low bioavailability (50-60%) following oral administration. Metformin is a substrate of organic cation transporters 1, 2 and 3 (OCT1-3) and its oral absorption, hepatic uptake and renal excretion occur mainly via these transporters. Membrane monoamine transporter (PMAT) expressed in human intestine may play a role in the intestinal absorption of metformin. Additionally, metformin is a good substrate for human multidrug and toxin extrusion 1 (MATE1) and MATE2-K. Metformin displays flip–flop kinetics (in human $t_{1/2,po}/t_{1/2,iv}$ =4.1) after oral administration. Its binding to human plasma proteins is negligible after absorption. Metformin accumulates in the kidney, salivary glands and walls of the small intestine, and it has a large apparent volume of distribution. The average elimination half-life in plasma is about 6 hours. Metformin is not metabolized, and it is eliminated from the body by tubular secretion and excreted unchanged in the urine. Physicochemical, pharmacological, and pharmacokinetic properties of metformin will be reviewed this chapter.

Chapter 3 - Metformin is currently the most widely-used drug for the treatment of diabetes mellitus. Perhaps the main therapeutic effect of metformin (not only an unwanted one) is restriction of aerobic metabolism resulting in lower energy income and adenosine triphosphate (ATP) production. The ultimate effect is increased income of glucose to the cell enabling constant ATP production. This mechanism explains why metformin decreases glucose production by liver cells and restores its uptake by skeletal muscle cells from the blood, even if they do not respond sufficiently to insulin-stimulated impulses. Metformin increases intestinal absorption of glucose, but its turnover into lactate is significantly increased, with intestinal metabolism subsequently becoming ineffective (more anaerobic). The rate of glucose absorption is reduced, and the process occurs more distally along the tract. In the end, metformin decreases net intestinal glucose uptake and aerobic glucose utilization. However, the side effects of this drug can be dangerous. Metformin-induced lactic acidosis (MILA) is a lethal disorder. The first significant deviation – and probably the cause of MILA development in monitored patients – is gastrointestinal tract involvement. Common infection or the effect of metformin itself cause vomiting and diarrhea, with subsequent hypovolemia causing pre-renal acute kidney injury, which in turn raises metformin levels up to toxic values. Development of kidney failure (followed by heart failure with dyspnea and brain failure with a delirious state) is often unexpected and fast. Metformin has become the most widely-used drug for the treatment of Type-2 diabetes mellitus. Strictly speaking – we use this drug in cases of treatment of "syndrome of intracellular overfeeding with secondary syndrome of chronic hyperglycemia", today known as "DMT2 syndrome", or "Type-2 diabetes mellitus with obesity". It is the drug of choice for the treatment of T2DM in obese people today. Its use is not recommended in cases of DM with intracellular starvation, i.e., true "Type-1 diabetes mellitus, T1DM (a disease characterized by lack of insulin), but can be useful in patients with either type of diabetes mellitus, i.e., both T1DM and T2DM. Many patients with intracellular starvation due to a lack of insulin (T1DM) tend to overfeed – then treated with high, supra-normal, doses of insulin subsequently develop intracellular overfeeding and, hence, T2DM. On the other hand, many obese patients with T2DM lose their ability of insulin overproduction and they develop *either* type of diabetes mellitus. Today, metformin belongs to the safest medications in its class. Nevertheless – if we are forced to use drugs that limit the surplus of energy and decrease the effectiveness of metabolism – we can still expect an adverse effect characterized by aerobic metabolism collapse, i.e., MILA. Today we have some other modern drugs: incretins acting by the anorectic effect; gliflozins modulating direct glucose elimination via the urine. While adequate food intake and physical training have the same beneficial effect, they are definitely not associated with any potential harm.

Chapter 4 - Metformin remains the first choice and the most prescribed medicine for the treatment of diabetes. This pharmacological drug is effective in reducing glucose levels, has a low cost and causes mild side effects in a few numbers of patients. The reason why diabetic patients are prone to develop cancer is the combination of insulin resistance, high insulin circulating levels and increased insulin-like growth factors, all of them are well known in promoting and sustaining the growth of various type of tumours. In fact, metformin has other effects than decreasing both the insulin and IGF levels. It also acts in killing the tumour cells and consequently decreasing the tumour mass. A large number of studies both *in vitro* and *in vivo* have demonstrated that metformin treatment can lead tumour cells to anti-proliferative, autophagy and/or apoptosis processes. Since cancer induces cachexia state, which leads to an involuntary body mass waste, mainly muscle protein, we have seen previously that metformin could also recover some changed parameters on protein and glucose metabolism in an experimental model of cachexia, improving the host responses in these tumour-bearing rats. In this present chapter, the authors are going to explore the differences between healthy and cancer cells and how this can influence the metformin effect on these cells.

Chapter 5 - Metformin is the first-line drug used in the treatment of type 2 diabetes with more than six decades of clinical use. It is a relatively safe drug, with known pharmacokinetics and a favorable safety profile. In recent years, in addition to hypoglycemic effect, several epidemiological studies have highlighted the therapeutic potential of metformin in the context of other diseases, especially in breast and ovarian cancers; thus, demonstrating its beneficial actions on survival outcome and prevention of these highly lethal malignancies. Recent and robust data support the fact that metformin would provide invaluable benefits when administrated along with conventional adjuvant therapy against the breast and ovarian cancers. Metformin impairs cellular metabolism, suppresses oncogenic signaling pathways, inhibits cell growth and metastasis, as well as reverses chemoresistance. Thus, improving treatment efficacy and allowing the prescription of lower doses of highly toxic antineoplastic agents. Altogether, the aforementioned information points out metformin as an anticancer agent. Of clinical interest, it arises as a therapeutic possibility in the fight against poor prognosis diseases, such as triple-negative breast cancer and high grade serous ovarian cancer. In the present chapter, the authors summarized the molecular mechanisms underlying the anticancer effects of metformin in breast and ovarian cancers. Furthermore, the authors discuss the potential anticancer applications of metformin-based combinatorial chemotherapy.

In: The Pharmacological Guide to Metformin ISBN: 978-1-53616-634-7
Editor: Natalie Lajoie

Chapter 1

THE PHARMACOGENETICS OF METFORMIN

***Gerard Marshall Raj*[1,*] *and Saranya Vilvanathan*[2,#]**
[1]Sri Venkateshwaraa Medical College Hospital and Research Centre, Puducherry, India
[2]Medical University of the Americas, Charlestown, Nevis, Saint Kitts and Nevis, West Indies

ABSTRACT

After almost six decades of its discovery, metformin still remains the gold standard drug for the management of type 2 diabetes mellitus owing to its euglycemic property and cost-effectiveness. It is the most commonly prescribed and utilized oral antidiabetic drug worldwide. However, no two persons with the same phenotypical characteristics, such as age, gender, body weight and others, tend to respond similarly to equivalent doses of metformin. Moreover, a higher proportion of patients fail to attain optimal glycemic control with metformin. This variability in glycemic response to metformin could be attributed to the differences in the genetic architecture of the individuals. Various pharmacokinetic and

* Corresponding Author's E-mail: drgmr1111@gmail.com.
# E-mail: saranyavilvan@gmail.com.

pharmacodynamic genetic markers determine the ultimate glycemic response to metformin.

Metformin is a charged hydrophilic molecule and it relies heavily on the membrane transporters to traverse across the biological membranes. It is not metabolized inside the human body and is excreted unaltered through the liver and kidneys. Hence, the optimal way to assess the pharmacokinetic involvement is to study the genes coding for the metformin drug transporters, which are involved in the absorption, distribution and elimination of the drug. The transporters of metformin include both the influx (which help the metformin to enter the cells) and efflux (which push the metformin outside the cells) transporters located at various cellular sites, namely, the intestine, the liver and the kidneys. The influx transporters include the OCT1 (organic cation transporter 1), OCT2, OCT3 and PMAT (plasma membrane monoamine transporter); the efflux transporters include MATE1 (multidrug and toxin extrusion protein 1) and MATE2. All these transporters are secondary active transporters and belong to the solute carrier (SLC) superfamily of transporters; the secondary active transporters obtain energy from the oppositely directed proton gradient, hence, they are also known as the H^+/organic cation antiporters. The OCTs are polyfunctional transporters which aid in the influx of metformin into cells. The OCT1 is located predominantly at the sinusoidal surface of the liver cells and also at the basolateral membrane of enterocytes; whereas, the OCT2 is expressed basolaterally in the kidney tubular cells. The OCT3 and PMAT are present at the luminal (apical) side of the intestinal cells. The metformin efflux transporters MATE1 and MATE2 are located in the luminal (canalicular) side of the liver and kidney, respectively; the MATE1 transporters are additionally present in the luminal side of the kidney. The facilitative bidirectional glucose transporter GLUT2 located in the liver, kidney, pancreas and small intestine is also implicated as one of the major genetic determinants of metformin response in large genome-wide association studies. The genes coding for OCT1, OCT2, OCT3, PMAT, MATE1, MATE2 and GLUT2 are *SLC22A1*, *SLC22A2*, *SLC22A3*, *SLC29A4*, *SLC47A1*, *SLC47A2* and *SLC2A2,* respectively.

Studies done to assess the pharmacodynamic aspect of metformin, to predict the therapeutic efficacy and safety and to gain insight into its mechanism of action have shown that there are various potential genetic factors responsible for the variable therapeutic response of the drug. The most commonly implicated among these is *ATM* gene polymorphisms (ATM stands for *ataxia telangiectasia*); *ATM* gene encodes serine-threonine kinases that are involved in cell-cycle regulation and in the phosphorylation of AMP-kinase (AMPK) which is a major target for metformin. Polymorphisms in *ATM* gene can alter the AMPK levels which in turn can affect the glycemic response to metformin.

Though the field of pharmacogenetics of metformin is amassing more and more wealth of knowledge the actual translation of the same to the clinical settings is suboptimal; the reasons being the multigenetic involvement and lack of adequately powered and scientifically robust studies. In this chapter, we focus on the various genetic polymorphisms that influence the therapeutic effects of metformin.

1. INTRODUCTION

Diabetes mellitus is a growing epidemic accounting for 425 million patients worldwide as of 2017; among these 90% are type 2 diabetes mellitus (T2DM) cases. In 2045, the number of people living with diabetes may rise by 48% to 629 million. The low- and middle-income countries account for 79% of the diabetes population [1]. Furthermore, diabetes is considered to be a major contributing factor in increasing the susceptibility to infections in these areas [2]. The predictions of the global economic burden studied across 180 countries are also ominous with an increase in costs as a share of global gross domestic product (GDP) from 1.8% (1.7 to 1.9) in 2015 to a maximum of 2.2% (2.1 to 2.2) in 2030 [3].

Diabetes is one among the leading causes for end-stage renal disease – 10 times higher incidence, cardiovascular disease – 2 to 3 times higher incidence, neuropathy and retinopathy – the leading cause of blindness [1]. Moreover, diabetes is also associated with high mortality rates with around 1.37 million deaths due to diabetes globally in 2017 [4].

Apart from supplementary lifestyle modifications, effective pharmacotherapeutic measures are mandatory. Metformin is the most commonly prescribed and utilized oral antidiabetic drug worldwide due to its euglycemic property and cost-effectiveness. It remains as the gold standard and more often the first drug of choice for the management of T2DM according to the current treatment guidelines recommended by various professional bodies, including the American Diabetes Association (ADA) [5], the International Diabetes Federation (IDF) [6], the European Association for the Study of Diabetes (EASD) [7], the American Association of Clinical Endocrinologists and the American College of

Endocrinology (AACE & ACE) [8] and the National Institute for Health and Care Excellence (NICE) [9].

2. Variable Response to Metformin

Not all patients treated with metformin respond similarly; it has been found that nearly 30% of the patients do not respond optimally to metformin monotherapy. Both genetic and non-genetic phenotypic factors influence this variability in glycemic response to metformin. However, the phenotypic features such as age, sex and body weight are only partly responsible for the differential response to metformin [10–12].

Hence, the heterogeneous genetic structure among individuals can be studied to assess the inter-patient variability to metformin response. The genes which code for the various proteins in the pharmacokinetics and pharmacodynamics pathway are studied elaborately to ascertain the influence of genetic variations on glycemic response [13–15].

3. Genetic Influence on Pharmacokinetics of Metformin

Metformin is an organic hydrophilic cation and hence, relies on membrane transporters to cross the phospholipid bilayers. With the aid of organic cation transporters, metformin is extensively distributed into the intestine, liver and kidney. Metformin does not undergo chemical alteration in the body and is excreted unchanged in the urine through active tubular secretion in the kidney. Hence, the genetic variations in these transporters are major contributors to the pharmacokinetics of metformin and the consequent glycemic response [16].

Metformin is transported across cell membranes by influx and efflux transporters. OCT1 (organic cation transporter 1), OCT2, OCT3 and PMAT (plasma membrane monoamine transporter) are the influx transporters which help in transporting metformin into the cells. Correspondingly, MATE1 (multidrug and toxin extrusion protein 1) and MATE2 are the major efflux transporters that export metformin outside the cells. These transporters belong to the solute carrier (SLC) superfamily of transporters and are termed as *secondary active transporters* – as they derive energy from the oppositely directed proton gradient. Hence, they are also known as the H^+/organic cation antiporters.

The OCT1 is located predominantly at the sinusoidal surface of the liver cells and also at the basolateral membrane of intestinal cells (enterocytes) and the OCT2 is expressed basolaterally in the kidney tubular cells. The OCT3 and PMAT are present at the luminal (apical) side of the intestinal cells.

The metformin efflux transporters MATE1 and MATE2 are located in the luminal (canalicular) side of the liver and kidney, respectively; the MATE1 transporters are also present in the luminal side of the kidney.

GLUT2 is a facilitative bidirectional glucose transporter located in the liver, kidney, pancreas and small intestine.

The genes coding for OCT1, OCT2, OCT3, PMAT, MATE1, MATE2 and GLUT2 are *SLC22A1*, *SLC22A2*, *SLC22A3*, *SLC29A4*, *SLC47A1*, *SLC47A2* and *SLC2A2*, respectively [16, 17]. Genetic polymorphisms in these genes are found to influence therapeutic response to metformin.

3.1. Organic Cation Transporters (OCTs)

Though equivocal, candidate gene studies in *SLC22A1* and other metformin transporters have been found to influence the pharmacokinetics, therapeutic response and gastrointestinal adverse effects to metformin [18]. A recent comprehensive review, by Mannino et al. [19], on pharmacogenetics of T2DM identified 64 genes and around 200 informative genetic variants associated with the response to oral antidiabetic drugs. They also delineated the importance of association of OCT family, *ATM* and *SLC2A2* genes with metformin response.

3.1.1. Organic Cation Transporter 1 (OCT1, Encoded by SLC22A1)

OCT1 is encoded by *the SLC22A1* gene, located in the long arm of chromosome 6q25.3 and consists of 11 exons spanning 37kb. OCT1 contains 554 amino acids with 12 transmembrane domains [20]. *In vitro* studies have demonstrated that apart from various exogenous and endogenous factors that regulate the expression of OCT1, genetic polymorphisms in *SLC22A1* can also contribute to the differential expression. This, in turn, can contribute to the inter-individual variation in the disposition and distribution of response and substrate drugs [21, 22]. Shu et al. [23] studied the correlation between polymorphisms in *OCT1* gene and metformin response in *in vitro* assays, knock-out mice as well as in healthy volunteers. *In vitro* assays confirmed that OCT activity is determinant of metformin response. Out of 12 non-synonymous variants, 7 *OCT1* variants (R61C, G401S, M420del, S14F, S189L, V408M and G465R) exhibited significantly reduced or lost metformin uptake; studies in *Oct1* knockout mouse showed reduced uptake of metformin and decreased response in hepatocytes; genotype-phenotype correlation in healthy volunteers showed that defective variants had significantly decreased glucose-lowering response to metformin. Shu et al. [24] also described the role of *OCT1* polymorphisms on metformin pharmacokinetics. *OCT1* genotypes carrying reduced function allele (R61C, G401S, 420del, G465R) exhibited a significant effect on metformin pharmacokinetics and were associated with higher AUC, maximum plasma concentration and lower oral volume of distribution.

However, in a subsequent Genetics of Diabetes Audit and Research Tayside Scotland (GoDartS) study, two common *SLC22A1* variants, R61C (rs12208357) and 420del (rs72552763) were reported to have no association with the impaired initial response to metformin, or metformin monotherapy failure [25].

Shikata et al. [26] also showed that the polymorphisms in *OCT1* (-43T>C) and *OCT2* (Val408Met) genes contribute to the clinical efficacy of metformin. Tzvetkov et al. [21] studied the polymorphisms in the transporter genes, including *OCT1*, *OCT2*, *OCT3*, *OCTN1* and *MATE1* polymorphisms, on the renal clearance of metformin. The study concluded that the renal clearance of metformin showed substantial interindividual variability which was reflected in the bioavailability. Genetic variation in *OCT1* affected the renal clearance of metformin but the common variants in *OCT2* and *OCT3* did not have any effect. Becker et al. [27] studied 11 tagging single nucleotide polymorphisms (SNPs) in *SLC22A1* gene and their association with the glucose-lowering response of metformin. He concluded that the SNP rs622342 A>C is associated with HbA_{1c} reduction. For each variant "C" allele, reduction in HbA_{1c} was found to be 0.28%. Becker et al. [28] went on to study the interaction between polymorphisms in *OCT1* and *MATE1* on metformin response and found that an interaction exists between the *OCT1* rs622342 A>C and *MATE1* rs2289669 G>A polymorphisms with respect to HbA_{1c} reduction. The decrease in HbA_{1c} was greater for wild "A" allele of rs622342 in *OCT1* and variant "A" allele of rs2289669 in *MATE1* genes. In "AA" homozygotes of both the alleles the reduction in HbA_{1c} was found to be the largest (-0.91%) whereas variant "CC" genotype of *OCT1* and wild "GG" genotype of *MATE1* showed an increase in HbA_{1c} by 0.48%. Christensen et al. [29] evaluated the effect of genetic variations in OCT1, OCT2, MATE1, MATE2 and PMAT on trough steady-state plasma concentration of metformin and HbA_{1c} as a part of the South Danish Diabetes Study (a 2 x 2 x 2 factorial, prospective, randomized, double-blind, placebo-controlled, multicentre study) comprising 159 patients. The study concluded that there was around 80-fold variability in the trough steady-state plasma concentration of metformin and OCT1 activity affects the metformin steady-state pharmacokinetics. It was also found that loss-of-function mutation in *OCT1* gene is associated with diminished pharmacodynamic (HbA_{1c}) effect of metformin after 6 and 24 months. Gambineri et al. [30] showed that the four variants of *OCT1* genotype (R61C, C>T; G401S, G>A; G465R, G>A and 429del) in patients with polycystic ovarian disease (PCOS) is a determinant of lipid and insulin response of metformin in addition to its effect on hepatocytes. Ahlin et al. [31] studied the genotype-dependent effects of drug interaction between OCT1 inhibitors and metformin. In the *in vitro* analysis, it was found that among the 4 *OCT1* variants studied (R61C, V408M, M420del and G465R), two variants (M420del and R61C) were found to be more sensitive to drug inhibition with verapamil and amitriptyline. M420del produced a larger decrease in transport of metformin in the presence of inhibitors which could be reflected in the plasma concentrations of metformin. The association of two loss-of-function variants of *SLC22A1*, namely, R61C (rs12208357) and M420del (rs72552763) with the incidence of gastrointestinal side effects of metformin was reported by Dujic T et al. [32] in a prospective observational cohort study. *SLC22A*1 (rs594709) was found to influence the metformin efficacy in Chinese T2DM population (n = 267) in a study Xiao et al. [33]. In Iranian diabetic population, the SNP rs628031 in *SLC22A1* gene was studied and it was found that there was no significant association of the polymorphism with metformin response [34]. Dawed et al. under the Innovative Medicines Initiative: DIabetes REsearCh on patient stratification (IMI DIRECT) consortium, has found a strong association of SNP rs3889348 in *SLC22A1* with the incidence of gastrointestinal intolerance in metformin-treated diabetic patients [35]. The genotyping of rs622342 in *SLC22A1* has been useful in predicting the response to metformin in combination therapy in Egyptian and South Indian T2DM patients [36, 37]. Among the important SNPs studied in *SLC22A1* gene, as per the NCBI database, the rs622342 SNP has been more strongly associated with the metformin response, followed by rs1867351, rs2297374 and rs4646272.

Apart from metformin-related pharmacogenetics, the importance of the OCT1 transporter in drug development was recently emphasized by the International Transporter Consortium (ITC). The ITC included the polymorphism in *SLC22A1* gene (OCT1) along with the polymorphisms in *ABCG2* (BCRP) and *SLCO1B1* (OATP1B1) genes as the major determinants of interindividual differences in the plasma levels, adverse effects and therapeutic response to many drugs [38].

3.1.2. Organic Cation Transporter 2 (OCT2, Encoded by SLC22A2)

OCT2 is predominantly distributed in the renal tubular cells and is found to be significantly associated with the elimination of metformin, which in turn can affect the disposition of metformin [22, 39]. *SLC22A2* gene is found to be highly polymorphic [22] and there are several studies that associate the loss-of-function mutation in *SLC22A2* with the disposition of metformin. Wang et al. [22] characterized the genetic variations of *OCT2* gene in Chinese population and investigated the potential functional significance of *OCT2* polymorphism using metformin with or without cimetidine. A total of 14 genetic variants were identified and 13 had frequency >1%. The polymorphism 808G>T is associated with reduced metformin renal or tubular clearance. This mutation also affects the inhibitory effect of cimetidine on the renal tubular secretion of metformin. Shikata et al. [26] reported a little contribution of *OCT2* polymorphism in the clinical efficacy of metformin. Chen et al. [40] studied the effect of *OCT2* [808G>T] polymorphism on pharmacokinetics of metformin in *in vitro* studies as well as in healthy volunteers of Caucasian and African American descent. In the *in vitro* assay, the *OCT2*-808T variant had a greater capacity to transport metformin. The variant allele in *OCT2*, among the healthy volunteers, was found to play a role in the renal clearance and net secretion of metformin. Song et al. [41] evaluated the contribution of OCT2 in renal tubular transport of metformin and studied the effects of 3 non-synonymous genetic variants in *OCT2* on transport activity of metformin in oocytes. He reported that in the renal tubular cells, the basolateral to apical transport of metformin is higher than the apical to basolateral transport. Genetic variants in OCT2 significantly decreased the transport activity of metformin which contributes to the inter-individual variation. Li et al. [42] reported that 808G>T variance in the *SLC22A2* gene can affect the plasma lactate level and the incidence of hyperlacticaemia in T2DM patients undergoing metformin therapy. The female patients carrying the "TT" (variant) genotype are prone to lactatemia. Zaharenko et al. [43] identified the association between the lack of metformin response and SNPs rs3119309 and rs7757336 located in the 5' flanking region of the genes coding for *OCT2* in 233 newly-diagnosed TDM Slovakians. A randomized cohort study in T2DM patients with one-year follow-up demonstrated that the efficacy of metformin is influenced by rs316019 variant of *SLC22A2* [44]. The SNP rs201919874 in *SLC22A2* was significantly associated with the pharmacokinetics of metformin in a cohort of Pakistani diabetes patients (n = 800) [45].

The vectorial renal elimination of organic cationic drugs, including that of metformin, is possible by the coordinated function of the influx OCT2 and the efflux MATE2 transporters [46]. Hence, the net renal elimination of any organic cation is a composite function of influx and efflux transporters. Christensen et al. [47] studied the gene-gene interaction between polymorphisms in the *SLC22A2* and *SLC47A1* genes and propounded the counteracting effects of the rs316019 and rs2252281 variants on the renal elimination of metformin.

3.1.3. Organic Cation Transporter 3 (OCT3, Encoded by SLC22A3)

Experimental [48, 49] and clinical studies [50] have clarified the pharmacokinetic significance of OCT3, expressed in the brush border membrane of the enterocytes, in determining the bioavailability of metformin. *In vitro* studies have also suggested the importance of missense variants in the *SLC22A3* gene [T44M (c.131C > T), T400I (c.1199C > T) and V423F (c.1267G > T)] on the therapeutic effects of metformin [51].

Recently, the pharmacokinetic relevance of two major promotor haplotypes of OCT3 was deciphered in the Korean population [52].

A study in the Iranian population (n = 150) revealed lack of association between the SNP rs3088442 in the 3´-untranslated region (-564G>A) of the *SLC22A3* gene with the glycemic control [53]; on the other hand the presence of the rs3088442G>A variant was confirmed as a genetic marker for the susceptibility to T2DM in the same population [54]. However, a synonymous variant rs2292334 (-1233G > A) of the *SLC22A3* gene was found to influence the therapeutic response to metformin; in the patients with the “A” allele (0.77% reduction from baseline) the mean reduction in HbA_{1c} was higher than those with the homozygous “G” allele (0.54% reduction from baseline) [55].

In a study with Pakistani T2DM patients (n = 600), the overexpression of the “A” allele of the rs3088442 G >A variant was associated with better clinical response to metformin [56].

Another variant, rs12194182 was found to have a significant association ($p < 0.05$) with lower mean HbA_{1c} levels ("CC" genotype group) in the Jordanian population (n = 212). However, after correction for multiple testing, the association was negated. Nevertheless, along with the other genotypes of *SLC22A1*, *SLC22A2* and phenotypic characteristics such as body mass index (BMI) and age of diagnosis, the aforementioned SNP was found to be associated with glycemic response to metformin following multinomial logistic regression ($p < 0.05$) [57].

An SNP (rs2481030) located in the 5' flanking region of the *SLC22A3* gene was found to be associated with reduced plasma concentrations of metformin in Slovakian diabetic patients; the results were also replicated in a small cohort of 25 healthy volunteers [43].

A clinical epigenetic study revealed that metformin decreases DNA methylation of metformin transporter genes in the human liver. Particularly, DNA methylation of the *SLC22A3* gene was related to lower expression of *SLC22A3* gene and higher fasting glucose levels in patients on insulin plus metformin treatment [58].

The OCT3 transporters are also expressed both at the basolateral and apical membranes of salivary gland acinar cells. Hence, they are also found to be responsible for accumulation and secretion of metformin from the salivary glands and the consequential, metformin-induced altered taste [59].

3.2. Multidrug and Toxin Extrusion Proteins (MATEs)

Functional alteration of transport activities of MATE proteins consequent to genetic variation in the MATE genes has been known [60]. The MATEs transporters were identified lately, i.e., 15 years back, and are gaining clinical relevance with respect to renal and hepatic organic cation excretion – including that of metformin [61].

3.2.1. Multidrug and Toxin Extrusion Protein 1 (MATE1, Encoded by SLC47A1)

The therapeutic relevance of MATE1 in metformin pharmacokinetics was further strengthened by co-administration of famotidine (a selective MATE1 inhibitor) with metformin which resulted in enhanced bioavailability and accentuated renal clearance of metformin [62]. *In vitro* analysis has also proven the importance of MATE1 – like the *Mate1* knockout homozygous mice (-/-) were found to be more prone for lactic acidosis due to increased metformin concentration [63].

Becker et al. [64] conducted the maiden metformin-based pharmacogenetic study on MATE1 transporters and identified that the SNP rs2289669 was significantly associated with the HbA_{1c} in 116 T2DM patients (Caucasians) on metformin. It was concluded that the glucose-lowering effect of metformin was higher in subjects with the "AA" genotype compared to other genotypes; the HbA_{1c} reduction was 0.30% higher for each minor "A" present ($p = 0.045$). Choi et al. [65] in their study population (n = 189, Caucasians) identified similar association of the variant rs2289669 with relative change in HbA_{1c} ($p = 0.055$). Tkáč et al. [66] reported similar findings in another group of Caucasian metformin-naïve T2DM patients (n = 148). The homozygous "A" allele carriers of the SNP rs2289669 showed a two-fold (0.55%) greater reduction of HbA_{1c} than those with the "GG" and "AG" genotypes ($p = 0.018$) after 6 months of metformin therapy. He et al. [67] conducted a similar study in newly diagnosed T2DM patients of the Chinese population (n = 220). As with other previous studies, it was confirmed that the patients who had the "AA" genotype responded significantly better than the carriers of other genotypes ("AG" and "GG" genotypes). The reduction in HbA_{1c} in patients with "AA", "GA" and "GG" genotype groups were -2.32%, -1.16% and -1.07% ($p < 0.05$), respectively, after 1 year of metformin treatment. In this study, it was also found that the *SLC47A1* genotype was an independent determinant for urine excretion of metformin ($p < 0.01$) after multivariate linear regression. Recently, Liang et al. [68] and Mousavi et al. [69] also described that the T2DM patients possessing the minor "A" allele exhibited superior glycemic response to metformin than those with the "G" allele after evaluating the Chinese (n = 291) and Iranian (n = 71) population, respectively. Liang et al. [68] also extrapolated that the inter-individual variability in response to metformin could be due to the differential increases in basal GLP-1 concentrations across genotypes.

Stocker et al. [70] reported that the Caucasian diabetic patients (n = 145) with rs2252281 variant (g.–66T→C) in *SLC47A1* gene showed heightened metformin response.

However, some of the studies reported a lack of association between the SNP rs2289669 and glycemic response to metformin. Klen et al. [71] experimented on 135 Slovenian T2DM patients on metformin for 6 months and concluded that there was no association between the rs2289669 variant and the HbA_{1c} levels ($p = 0.222$); however, they found a positive link between the same SNP and total cholesterol levels ($p = 0.018$). Tarasova et al. [72] also reported no association between this particular variant and glycemic response or adverse effects due to metformin. A corrigendum published by Christensen et al. [73] also stated that there was no significant association between this SNP and the decrease in HbA1c, upending their previous findings. Xiao et al. [33] also proved the lack of direct association between the SNP rs2289669 and glycemic response to metformin monotherapy; however, among *SLC22A1* rs594709 "AA" genotype patients who also possessed *SLC47A1* rs2289669 "AA" genotype the decrease in FBG ($p = 0.015$) was higher than "G" allele carriers.

In the South Indian population, Raj et al. [74] reported lack of association of the SNP rs2289669 with the glycemic response to metformin ($p = 0.079$) after three months of metformin monotherapy (n = 105). Another study by Phani et al. [75] in a different cohort of South Indian population (n = 188) also revealed similar results with both rs2289669 ($p > 0.05$) and rs8065082 ($p > 0.05$) variants after three months of metformin monotherapy.

A pharmacokinetic study (n = 26, healthy Koreans), by Cho et al. [76], also revealed that the SNP rs2289669 in *SLC47A1* gene was found to influence the renal clearance of metformin only after the treatment with ranitidine (a potential MATE inhibitor). *In vitro* studies have confirmed the clinical relevance [77] and lack of substrate identity [78] for MATE1-mediated drug-drug interactions.

3.2.2. Multidrug and Toxin Extrusion Protein 2 (MATE2, Encoded by SLC47A2)

As MATE2 is expressed at the renal tissues, it is also referred to as MATE2K. Choi et al. [65] had also studied the influence of four nonsynonymous variants and four variants in basal promoter regions of the *SLC47A2* gene in 253 T2DM patients (Caucasians: 189, African Americans: 64) on metformin monotherapy. The novel promoter variant g.−130G>A located in the 5'UTR region (rs12943590) was associated with poor response to metformin. They concluded that those patients who were homozygous carriers of the minor "A" allele (5.9% reduction) had a significantly poorer decrement (p = 0.031) in HbA_{1c} values than those patients with the "GG/AG" genotypes (14% reduction). Stocker et al. [70] had obtained similar results with variant carriers displaying reduced glycemic response ($p < 0.05$); this was due to increased renal clearance of metformin.

Raj et al. [74] showed that the patients with "GG" genotype had a 5.5% decrease in HbA_{1c} from baseline compared to those with the "AG/AA" genotype (0.1% increase). The above SNP in the 5′UTR of the *SLC47A2* gene was found to be in linkage disequilibrium ($r^2 > 0.80$) with the intronic SNPs like rs9900497 and rs35263947 and a synonymous exonic SNP rs4924792. Thus, the presence of these SNPs would also have contributed to the glycemic response exhibited. However, after adjusting for multiple comparisons (Bonferroni correction) the difference found in the glycemic response between the "GG" genotype and "AG/AA" genotype groups of the *SLC47A2* rs12943590 SNP was not significant (p = 0.027, which was greater than the critical value of 0.025). Conversely, Phani et al. [75] reported a positive association between the same SNP with metformin response (p = 0.035, co-dominant model). They also revealed that those with the *SLC22A2* rs316019 "GG" and *SLC47A2* rs12943590 "GA" combined genotypes had a maximum reduction in post-treatment HbA_{1c} levels (p = 0.02).

Moeez et al. [45] reported a positive association between the SNP rs138244461 in *SLC47A2* gene and pharmacokinetics of metformin after studying the Pakistani cohort.

Recently, *in vitro* studies have also proven that the expression of MATE2 influences the cancer cell response to metformin [79].

The clinical relevance of MATE2K was questioned by the study by Morrissey et al. [80] who divulged that selective inhibition of MATE2K (with nizatidine) did not change either the systemic concentrations or the renal clearance of metformin. Moreover, nonspecific inhibition of MATE transporter with pyrimethamine also was not found to be associated with the antihyperglycaemic activity of metformin [81]. In a small Japanese diabetic cohort (n = 48), it was found that neither the MATE1 nor the MATE2K was associated with metformin disposition [82].

A large-scale meta-analysis by the Metformin Genetics (MetGen) Consortium in T2DM patients on metformin monotherapy of European ancestry (n = 5836) reported that the variability in glycemic response to metformin is not due to the genetic polymorphisms in the transporter genes. Five transporters, namely, OCT1 (rs12208357; rs72552763; rs622342), OCT2 (rs316019), MATE1 (rs2289669; rs2252281), MATE2K (rs12943590) and OCTN1 (rs272893; rs1050152), with nine SNPs were studied [83].

3.3. Glucose Transporter 2 (GLUT2, Encoded by *SLC2A2*)

A three-stage genome-wide association study (GWAS), consisting of 13, 123 participants hailing from different ethnic origins (10, 557 were of European ancestry and 2, 566 were of non-European ancestry), was conducted by the MetGen Consortium. The results revealed that the patients out of 13 different SNPs analyzed initially, the SNP rs8192675 in the intron of *SLC2A2* was associated with glycemic response to metformin. Among the European cohort, the metformin-induced reduction in hemoglobin A_{1c} (HbA_{1c}) was 0.17% greater in "C" allele ($p = 6.6 \times 10^{-14}$). The obese patients with the "CC" genotype showed a 0.33% (3.6

mmol/mol) greater absolute HbA_{1c} reduction than those with the "TT" genotype [84].

Recently, Rathmann et al. [85] recruited 508 T2DM patients from the German Diabetes Study and evaluated the association between the change in fasting glucose with the *SLC2A2* (rs8192675) variant. The metformin monotherapy patients with "C" allele displayed a greater adjusted blood glucose reduction during the first year after diagnosis of diabetes than those with the "TT" genotype (6.3 mmol/l vs 3.9 mmol/l; genotype difference 2.4 mmol/l, $p = 0.02$).

Hence, as depicted by the multiethnic GWAS, the rs8192675 variant of the facilitated glucose transporter (GLUT2) could be considered as a robust biomarker for stratified pharmacotherapy with metformin. The clinical effect of the SNP equates to a daily intake of additional 250 mg of metformin – which is 26% of the average daily dose.

3.4. Plasma Membrane Monoamine Transporter (PMAT, Encoded by *SLC29A4*)

Compared to other organic transporters (OCT1, OCT2 and OCT3), the clinical relevance of PMAT in metformin transport was identified [86] and confirmed [87] lately.

A cluster of SNPs (rs2685753, $p = 0.027$; rs3889348, $p = 0.015$; rs4720572, $p = 0.029$; rs4299914, $p = 0.025$ and rs6971788, $p = 0.047$) in the intronic region of *SLC29A4* was associated with decreased metformin absorption by the Christensen et al. study [29].

Under the IMI DIRECT consortium, Dawed et al. [35] reported that those patients with the "G" allele of rs3889348 variant were more prone for metformin-induced gastrointestinal intolerance (OR: 1.34, $p = 0.005$). Likewise, patients who possessed three or more risk alleles (including both *SLC29A4* and *SLC22A1* variants) the odds of intolerance were two times higher. Large scale studies like these are necessary to unravel the role of PMAT in the disposition and therapeutic response of metformin [88].

Recently, it has been divulged that the PMAT is highly expressed (PMAT >> OCT3 = MATE1) in the human fat cells and could be considered as the primary target for the antilipolytic effect of metformin [89].

None of the above discussed metformin transporters (with their SNPs) was found to influence the pharmacokinetic variability of metformin in a simulated population pharmacokinetic study in T2DM patients [90].

4. Genetic Influence on Pharmacodynamics of Metformin

Though physiologically, metformin has been found to reduce hepatic glucose output the molecular mechanisms include both AMP-activated protein kinase (AMPK)-dependent and AMPK-independent pathways; subsequent mechanisms include inhibition of mitochondrial respiration, inhibition of mitochondrial glycerophosphate dehydrogenase and also the involvement of lysosomes [91]. The intracellular concentration of metformin is a major determinant of the glycemic response to metformin [92].

4.1. ATM Protein Encoded by Ataxia Telangiectasia-Mutated (ATM) Gene

The ATM protein is a member of the phosphatidylinositol 3-kinase (PI-3K) family of proteins that respond to DNA damage by phosphorylating key substrates involved in DNA repair and/or cell cycle control. The influence of genetic polymorphisms on the pharmacodynamics, especially the mechanism of action, of metformin is evident from the studies associating the *ATM* gene polymorphism with metformin response. Though the mechanism of action of metformin is not completely understood, it is thought to inhibit the complex 1 of the respiratory chain in the mitochondria of liver cells thereby creating an oxidative stress and activation of AMPK. Interestingly, polymorphisms in the ATM gene were also found to affect the AMPK levels and in turn the glycemic response to metformin. The first GWAS on patients with T2DM was done as a part of the Wellcome Trust Case Control Consortium 2 study (WTCCC2), the GoDartS study and the United Kingdom Prospective Diabetes Study (UKPDS) to gain insight into the mechanism of action of metformin and to identify the potential genetic factors that can influence the efficacy and safety of metformin [93]. They identified 14 polymorphisms in a locus containing *ATM* gene that were associated with the ability to reach the treatment goal of HbA_{1c} <7%. Among these SNPs, rs11212617 was found to be strongly associated with the influence of metformin response. The carriers of minor "C" allele (frequency of 44%) had a higher rate of achieving the glycemic goal which was later replicated in two replication cohorts. This hypothesis was contraindicated [94, 95] with the fact that the compound KU-55933 used as ATM inhibitor in GWAS study also inhibited OCT1 transporters in hepatocytes and hence, inhibit the metformin uptake and consequently led to decrease in response. The authors also confirmed that the OCT1-mediated decrease in metformin uptake is independent of ATM and this gene does not have any detectable effect on OCT1 activity. Florez et al. [96] genotyped the SNP rs11212617 in *ATM* gene as a part of the Diabetes Prevention Program (DPP) and concluded that in contrast to the findings in GWAS, the "C" allele in *ATM* gene conferred no advantage on diabetes prevention among metformin users. In this study, metformin response was considered in terms of incidence of diabetes rather than reduction in HbA_{1c} values whereas GWAS study was done based on retrospective evaluation of clinical records. Van Leeuwan et al. [97] replicated the association established in the GWAS study in 3 different cohorts predominantly of Caucasian origin, including metformin users from Diabetes Care System (DCS); Rotterdam study, Netherlands and CARDS trial, UK. The association with treatment success was replicated in the first cohort (DCS) but not in the latter two. A meta-analysis was subsequently performed by the authors including the 3 cohorts along with the previously described GoDartS and UKPDS stage 2 replication cohorts, which confirmed that the minor "C" allele of rs11212617 had a higher therapeutic response to metformin in the Caucasian population. This hypothesis was strengthened by Tkac [98] who concluded that this SNP might belong to the panel of SNPs evaluated for metformin treatment in the future; however, this was not later replicated in the Iranian population [34].

In a Chinese T2DM population (n = 274), the clinical relevance of the SNP rs11212617 in *ATM* gene with treatment success was confirmed [99].

4.2. *PRKAA1*, *PRKAA2* and *STK11* Genes

Among the candidate genes that associate the incidence of diabetes with the response to metformin or lifestyle interventions, the highlighted ones, that are linked to the pharmacodynamics of metformin are those associated to the AMPK gene *STK11* and the AMPK subunit genes, *PRKAA1* and *PRKAA2*. The *STK11* gene codes for serine/threonine kinase; and the *PRKAA1* and *PRKAA2* genes code for protein kinase, AMP-activated, catalytic, alpha-1 and for protein kinase, AMP-activated, catalytic, alpha-2, respectively.

Jablonski et al. [13] analyzed 1590 SNPs in 2994 DPP participants and found nominal interactions of these genes with metformin response. Of the 91 nominal associations with metformin intervention, the SNPs rs741765 in *STK11* gene, rs249429 in *PRKAA1* and rs9803799 in *PRKAA2* were found to have significant interaction. Further studies are needed to establish the role of these putative genes, which are known to be encoding additional proteins associated with AMPK-dependent inhibition of gluconeogenesis.

4.3. *MEF2A* and *MEF2D* Genes

The SNPs rs4424892 in *MEF2A* and rs6666307 in *MEF2D* were also found to have significant interaction with metformin intervention in the

DPP trial done by Jablonski et al. [13]. These genes are known to be regulated by the levels of AMPK and cause an increase in mRNA expression of GLUT4 in skeletal muscles and adipose tissue and hence, affect the insulin sensitivity and response to exercise [100].

The MEF2A and MEF2D genes code for the MADS box transcription enhancer factor 2, polypeptide A and MADS box transcription enhancer factor 2, polypeptide D, respectively.

CONCLUSION

The Meta-Analysis Gene-set Enrichment of variaNT Associations (MAGENTA), an enrichment analysis, performed by Dawed et al. [101] revealed that the *SLC22A1*, *SLC47A1*, *STK11*, *ATM*, *PRKAA2* and *SLC22A2* genes and *SLC47A2*, *SLC22A3* and *SLC29A4* genes are associated with clinical outcomes of metformin and pharmacokinetics or pharmacodynamics of metformin, respectively; the expression of *SLC2A2* and *MEF2A* genes is also affected by metformin. The analysis was conducted with the support of the Fast Automated Biomedical Literature Extraction (FABLE) tool which auto-extracted 5963 articles related to metformin-gene associations.

Another study confirmed the lack of association between diabetes-related genetic risk scores (GRSs) and HbA_{1c} response in people with T2DM initiating oral glucose-lowering treatment (metformin, n = 282 or sulfonylureas, n = 89) [102].

Currently, the realms of pharmacogenetics have spread to the second-line antidiabetic drugs including the dipeptidyl peptidase-4 (DPP-4) inhibitors, glucagon-like peptide-1 (GLP-1) receptor agonists and sodium-glucose cotransporter-2 (SGLT2) inhibitors [103, 104]. Further, exploring the genetic variations may pave way to a broader understanding of diabetes and novel targets for diabetes pharmacotherapy [105, 106].

The concept of 'multi-omics' approach by horizontal integration of genomics with *transcriptomics* (RNA sequencing from fasting whole blood), *proteomics* (targeted array in fasting plasma), *metabolomics* (targeted and untargeted assays in fasting plasma) and *microbiomics* (DNA isolation and deep sequencing in faecal samples) may further strengthen the path towards application of precision medicine in the management of diabetes; the IMI DIRECT consortium [107] is one such group that is working towards this goal.

REFERENCES

[1] International Diabetes Federation. *IDF Diabetes Atlas*, 8th ed. Brussels, Belgium: International Diabetes Federation [Internet]. 2017 [cited 2019 Jul 28]. Available at: http://www.diabetesatlas.org.Atlas-8e-Global-factsheet.pdf.

[2] Dunachie S, Chamnan P. The double burden of diabetes and global infection in low and middle-income countries. *Trans R Soc Trop Med Hyg* 2019;113:56–64.

[3] Bommer C, Sagalova V, Heesemann E, Manne-Goehler J, Atun R, Bärnighausen T, et al. Global Economic Burden of Diabetes in Adults: Projections From 2015 to 2030. *Diabetes Care* 2018;41:963–70.

[4] GBD 2017 Causes of Death Collaborators. Global, regional, and national age-sex-specific mortality for 282 causes of death in 195 countries and territories, 1980-2017: a systematic analysis for the Global Burden of Disease Study 2017. *Lancet Lond Engl* 2018;392:1736–88.

[5] American Diabetes Association. 9. Pharmacologic Approaches to Glycemic Treatment: Standards of Medical Care in Diabetes-2019. *Diabetes Care* 2019;42:S90–S102.

[6] International Diabetes Federation Guideline Development Group. Global guideline for type 2 diabetes. *Diabetes Res Clin Pract* 2014;104:1–52.

[7] Davies MJ, D'Alessio DA, Fradkin J, Kernan WN, Mathieu C, Mingrone G, et al. Management of hyperglycaemia in type 2 diabetes, 2018. A consensus report by the American Diabetes

Association (ADA) and the European Association for the Study of Diabetes (EASD). *Diabetologia* 2018;61:2461–98.

[8] Garber AJ, Abrahamson MJ, Barzilay JI, Blonde L, Bloomgarden ZT, Bush MA, et al. Consensus statement by the American Association of Clinical Endocrinologists and American College of Endocrinology on the comprehensive type 2 diabetes management algorithm - 2019 executive summary. *Endocr Pr Off J Am Coll Endocrinol Am Assoc Clin Endocrinol 2019*;25:69–100.

[9] Overview | *Type 2 diabetes in adults: management* | Guidance | NICE [Internet]. [cited 2019 Aug 7]. Available from: https://www.nice.org.uk/guidance/ng28.

[10] Cook MN, Girman CJ, Stein PP, Alexander CM. Initial monotherapy with either metformin or sulphonylureas often fails to achieve or maintain current glycaemic goals in patients with Type 2 diabetes in UK primary care. *Diabet Med J Br Diabet Assoc* 2007;24:350–8.

[11] Graham GG, Punt J, Arora M, Day RO, Doogue MP, Duong JK, et al. Clinical pharmacokinetics of metformin. *Clin Pharmacokinet* 2011;50:81–98.

[12] Han J, Yu H, Tu Y, Pang J, Liu F, Bao Y, et al. Different clinical prognostic factors are associated with improved glycaemic control: findings from MARCH randomized trial. *Diabet Med J Br Diabet Assoc* 2017;34:490–9.

[13] Jablonski KA, McAteer JB, de Bakker PIW, Franks PW, Pollin TI, Hanson RL, et al. Common variants in 40 genes assessed for diabetes incidence and response to metformin and lifestyle intervention in the diabetes prevention program. *Diabetes* 2010;59:2672–81.

[14] Zhou K, Donnelly L, Yang J, Li M, Deshmukh H, Van Zuydam N, et al. Heritability of variation in glycaemic response to metformin: a genome-wide complex trait analysis. *Lancet Diabetes Endocrinol* 2014;2:481–7.

[15] Pawlyk AC, Giacomini KM, McKeon C, Shuldiner AR, Florez JC. Metformin pharmacogenomics: current status and future directions. *Diabetes* 2014;63:2590–9.

[16] Liang X, Giacomini KM. Transporters Involved in Metformin Pharmacokinetics and Treatment Response. *J Pharm Sci* 2017; 106:2245–50.

[17] Gong L, Goswami S, Giacomini KM, Altman RB, Klein TE. Metformin pathways: pharmacokinetics and pharmacodynamics. *Pharmacogenet Genomics* 2012;22:820–7.

[18] Chan P, Shao L, Tomlinson B, Zhang Y, Liu Z-M. Metformin transporter pharmacogenomics: insights into drug disposition-where are we now? *Expert Opin Drug Metab Toxicol* 2018;14:1149–59.

[19] Mannino GC, Andreozzi F, Sesti G. Pharmacogenetics of type 2 diabetes mellitus, the route toward tailored medicine. *Diabetes Metab Res Rev* 2019;35:e3109.

[20] Jonker JW, Schinkel AH. Pharmacological and physiological functions of the polyspecific organic cation transporters: OCT1, 2, and 3 (SLC22A1-3). *J Pharmacol Exp Ther* 2004;308:2–9.

[21] Tzvetkov MV, Vormfelde SV, Balen D, Meineke I, Schmidt T, Sehrt D, et al. The effects of genetic polymorphisms in the organic cation transporters OCT1, OCT2, and OCT3 on the renal clearance of metformin. *Clin Pharmacol Ther* 2009;86:299–306.

[22] Wang Z-J, Yin OQP, Tomlinson B, Chow MSS. OCT2 polymorphisms and in-vivo renal functional consequence: studies with metformin and cimetidine. *Pharmacogenet Genomics* 2008; 18:637–45.

[23] Shu Y, Sheardown SA, Brown C, Owen RP, Zhang S, Castro RA, et al. Effect of genetic variation in the organic cation transporter 1 (OCT1) on metformin action. *J Clin Invest* 2007;117:1422–31.

[24] Shu Y, Brown C, Castro RA, Shi RJ, Lin ET, Owen RP, et al. Effect of genetic variation in the organic cation transporter 1, OCT1, on metformin pharmacokinetics. *Clin Pharmacol Ther* 2008;83:273–80.

[25] Zho K, Donnelly LA, Kimber CH, Donnan PT, Doney ASF, Leese G, et al. Reduced-function SLC22A1 polymorphisms encoding organic cation transporter 1 and glycemic response to metformin: a GoDARTS study. *Diabetes.* 2009 Jun;58(6):1434–9.

[26] Shikata E, Yamamoto R, Takane H, Shigemasa C, Ikeda T, Otsubo K, et al. Human organic cation transporter (OCT1 and OCT2) gene polymorphisms and therapeutic effects of metformin. *J Hum Genet* 2007;52:117–22.

[27] Becker ML, Visser LE, van Schaik RHN, Hofman A, Uitterlinden AG, Stricker BHC. Genetic variation in the organic cation transporter 1 is associated with metformin response in patients with diabetes mellitus. *Pharmacogenomics J* 2009;9:242–7.

[28] Becker ML, Visser LE, van Schaik RHN, Hofman A, Uitterlinden AG, Stricker BHC. Interaction between polymorphisms in the OCT1 and MATE1 transporter and metformin response. *Pharmacogenet Genomics* 2010;20:38–44.

[29] Christensen MMH, Brasch-Andersen C, Green H, Nielsen F, Damkier P, Beck-Nielsen H, et al. The pharmacogenetics of metformin and its impact on plasma metformin steady-state levels and glycosylated hemoglobin A1c. *Pharmacogenet Genomics* 2011;21:837–50.

[30] Gambineri A, Tomassoni F, Gasparini DI, Di Rocco A, Mantovani V, Pagotto U, et al. Organic cation transporter 1 polymorphisms predict the metabolic response to metformin in women with the polycystic ovary syndrome. *J Clin Endocrinol Metab* 2010;95:E204–208.

[31] Ahlin G, Chen L, Lazorova L, Chen Y, Ianculescu AG, Davis RL, et al. Genotype-dependent effects of inhibitors of the organic cation transporter, OCT1: predictions of metformin interactions. *Pharmacogenomics J* 2011;11:400–11.

[32] Dujic T, Causevic A, Bego T, Malenica M, Velija-Asimi Z, Pearson ER, et al. Organic cation transporter 1 variants and gastrointestinal side effects of metformin in patients with Type 2 diabetes. *Diabet Med J Br Diabet Assoc* 2016;33:511–4.

[33] Xiao D, Guo Y, Li X, Yin J-Y, Zheng W, Qiu X-W, et al. The Impacts of SLC22A1 rs594709 and SLC47A1 rs2289669 Polymorphisms on Metformin Therapeutic Efficacy in Chinese Type 2 Diabetes Patients. *Int J Endocrinol* 2016;2016:4350712.

[34] Shokri F, Ghaedi H, Ghafouri Fard S, Movafagh A, Abediankenari S, Mahrooz A, et al. Impact of ATM and SLC22A1 Polymorphisms on Therapeutic Response to Metformin in Iranian Diabetic Patients. *Int J Mol Cell Med* 2016;5:1–7.

[35] Dawed AY, Zhou K, van Leeuwen N, Mahajan A, Robertson N, Koivula R, et al. Variation in the Plasma Membrane Monoamine Transporter (PMAT) (Encoded by SLC29A4) and Organic Cation Transporter 1 (OCT1) (Encoded by SLC22A1) and Gastrointestinal Intolerance to Metformin in Type 2 Diabetes: An IMI DIRECT Study. *Diabetes Care* 2019;42:1027–33.

[36] Ebid A-HIM, Ehab M, Ismail A, Soror S, Mahmoud MA. The influence of SLC22A1 rs622342 and ABCC8 rs757110 genetic variants on the efficacy of metformin and glimepiride combination therapy in Egyptian patients with type 2 diabetes. *J Drug Assess* 2019;8:115–21.

[37] Umamaheswaran G, Praveen RG, Damodaran SE, Das AK, Adithan C. Influence of SLC22A1 rs622342 genetic polymorphism on metformin response in South Indian type 2 diabetes mellitus patients. *Clin Exp Med* 2015;15:511–7.

[38] Yee SW, Brackman DJ, Ennis EA, Sugiyama Y, Kamdem LK, Blanchard R, et al. Influence of Transporter Polymorphisms on Drug Disposition and Response: A Perspective From the International Transporter Consortium. *Clin Pharmacol Ther* 2018;104:803–17.

[39] Motohashi H, Sakurai Y, Saito H, Masuda S, Urakami Y, Goto M, et al. Gene expression levels and immunolocalization of organic ion transporters in the human kidney. *J Am Soc Nephrol Jasn* 2002;13:866–74.

[40] Chen Y, Li S, Brown C, Cheatham S, Castro RA, Leabman MK, et al. Effect of genetic variation in the organic cation transporter 2 on the renal elimination of metformin. *Pharmacogenet Genomics* 2009;19:497–504.

[41] Song IS, Shin HJ, Shin JG. Genetic variants of organic cation transporter 2 (OCT2) significantly reduce metformin uptake in

oocytes. *Xenobiotica Fate Foreign Compd Biol Syst* 2008;38:1252–62.

[42] Li Q, Liu F, Zheng T, Tang J, Lu H, Jia W. SLC22A2 gene 808 G/T variant is related to plasma lactate concentration in Chinese type 2 diabetics treated with metformin. *Acta Pharmacol Sin* 2010;31:184–90.

[43] Zaharenko L, Kalnina I, Geldnere K, Konrade I, Grinberga S, Židzik J, et al. Single nucleotide polymorphisms in the intergenic region between metformin transporter OCT2 and OCT3 coding genes are associated with short-term response to metformin monotherapy in type 2 diabetes mellitus patients. *Eur J Endocrinol* 2016;175:531–40.

[44] Hou W, Zhang D, Lu W, Zheng T, Wan L, Li Q, et al. Polymorphism of organic cation transporter 2 improves glucose-lowering effect of metformin via influencing its pharmacokinetics in Chinese type 2 diabetic patients. *Mol Diagn Ther* 2015;19:25–33.

[45] Moeez S, Khalid Z, Jalil F, Irfan M, Ismail M, Arif MA, et al. Effects of SLC22A2 (rs201919874) and SLC47A2 (rs138244461) genetic variants on Metformin Pharmacokinetics in Pakistani T2DM patients. *Jpma J Pak Med Assoc* 2019;69:155–63.

[46] Meyer zu Schwabedissen HE, Verstuyft C, Kroemer HK, Becquemont L, Kim RB. Human multidrug and toxin extrusion 1 (MATE1/SLC47A1) transporter: functional characterization, interaction with OCT2 (SLC22A2), and single nucleotide polymorphisms. *Am J Physiol Renal Physiol* 2010;298:F997–F1005.

[47] Christensen MMH, Pedersen RS, Stage TB, Brasch-Andersen C, Nielsen F, Damkier P, et al. A gene-gene interaction between polymorphisms in the OCT2 and MATE1 genes influences the renal clearance of metformin. *Pharmacogenet Genomics* 2013;23:526–34.

[48] Chen EC, Liang X, Yee SW, Geier EG, Stocker SL, Chen L, et al. Targeted disruption of organic cation transporter 3 attenuates the pharmacologic response to metformin. *Mol Pharmacol* 2015;88:75–83.

[49] Shirasaka Y, Lee N, Zha W, Wagner D, Wang J. Involvement of organic cation transporter 3 (Oct3/Slc22a3) in the bioavailability and

pharmacokinetics of antidiabetic metformin in mice. *Drug Metab Pharmacokinet* 2016;31:385–8.

[50] Hakooz N, Jarrar YB, Zihlif M, Imraish A, Hamed S, Arafat T. Effects of the genetic variants of organic cation transporters 1 and 3 on the pharmacokinetics of metformin in Jordanians. *Drug Metab Pers Ther* 2017;32:157–62.

[51] Chen L, Pawlikowski B, Schlessinger A, More SS, Stryke D, Johns SJ, et al. Role of organic cation transporter 3 (SLC22A3) and its missense variants in the pharmacologic action of metformin. *Pharmacogenet Genomics* 2010;20:687–99.

[52] Kwon EY, Chung J-Y, Park HJ, Kim BM, Kim M, Choi JH. OCT3 promoter haplotype is associated with metformin pharmacokinetics in Koreans. *Sci Reports* 2018;8:16965.

[53] Ghaffari-Cherati M, Mahrooz A, Hashemi-Soteh MB, Hosseyni-Talei SR, Alizadeh A, Nakhaei SM. Allele frequency and genotype distribution of a common variant in the 3´-untranslated region of the SLC22A3 gene in patients with type 2 diabetes: Association with response to metformin. *J Res Med Sci Off J Isfahan Univ Med Sci* 2016;21:92.

[54] Mahrooz A, Alizadeh A, Hashemi-Soteh MB, Ghaffari-Cherati M, Hosseyni-Talei SR. Polymorphic Variants rs3088442 and rs2292334 in the Organic Cation Transporter 3 (OCT3) Gene and Susceptibility Against Type 2 Diabetes: Role of their Interaction. *Arch Med Res* 2017;48:162–8.

[55] Hosseyni-Talei SR, Mahrooz A, Hashemi-Soteh MB, Ghaffari-Cherati M, Alizadeh A. Association between the synonymous variant organic cation transporter 3 (OCT3)-1233G>A and the glycemic response following metformin therapy in patients with type 2 diabetes. *Iran J Basic Med Sci* 2017;20:250–5.

[56] Moeez S, Riaz S, Masood N, Kanwal N, Arif MA, Niazi R, et al. Evaluation of the rs3088442 G>A SLC22A3 Gene Polymorphism and the Role of microRNA 147 in Groups of Adult Pakistani Populations With Type 2 Diabetes in Response to Metformin. *Can J Diabetes* 2019;43:128–135.e3.

[57] Al-Eitan LN, Almomani BA, Nassar AM, Elsaqa BZ, Saadeh NA. Metformin Pharmacogenetics: Effects of SLC22A1, SLC22A2, and SLC22A3 Polymorphisms on Glycemic Control and HbA1c Levels. *J Pers Med* 2019;9(1).

[58] García-Calzón S, Perfilyev A, Männistö V, de Mello VD, Nilsson E, Pihlajamäki J, et al. Diabetes medication associates with DNA methylation of metformin transporter genes in the human liver. *Clin Epigenetics* 2017;9:102.

[59] Lee N, Duan H, Hebert MF, Liang CJ, Rice KM, Wang J. Taste of a pill: organic cation transporter-3 (OCT3) mediates metformin accumulation and secretion in salivary glands. *J Biol Chem* 2014;289:27055–64.

[60] Kajiwara M, Terada T, Ogasawara K, Iwano J, Katsura T, Fukatsu A, et al. Identification of multidrug and toxin extrusion (MATE1 and MATE2-K) variants with complete loss of transport activity. *J Hum Genet* 2009;54:40–6.

[61] Nies AT, Damme K, Kruck S, Schaeffeler E, Schwab M. Structure and function of multidrug and toxin extrusion proteins (MATEs) and their relevance to drug therapy and personalized medicine. *Arch Toxicol* 2016;90:1555–84.

[62] Hibma JE, Zur AA, Castro RA, Wittwer MB, Keizer RJ, Yee SW, et al. The Effect of Famotidine, a MATE1-Selective Inhibitor, on the Pharmacokinetics and Pharmacodynamics of Metformin. *Clin Pharmacokinet* 2016;55:711–21.

[63] Toyama K, Yonezawa A, Masuda S, Osawa R, Hosokawa M, Fujimoto S, et al. Loss of multidrug and toxin extrusion 1 (MATE1) is associated with metformin-induced lactic acidosis. *Br J Pharmacol* 2012;166:1183–91.

[64] Becker ML, Visser LE, van Schaik RHN, Hofman A, Uitterlinden AG, Stricker BHC. Genetic variation in the multidrug and toxin extrusion 1 transporter protein influences the glucose-lowering effect of metformin in patients with diabetes: a preliminary study. *Diabetes* 2009;58:745–9.

[65] Choi JH, Yee SW, Ramirez AH, Morrissey KM, Jang GH, Joski PJ, et al. A common 5'-UTR variant in MATE2-K is associated with poor response to metformin. *Clin Pharmacol Ther* 2011;90:674–84.

[66] Tkáč I, Klimčáková L, Javorský M, Fabianová M, Schroner Z, Hermanová H, et al. Pharmacogenomic association between a variant in SLC47A1 gene and therapeutic response to metformin in type 2 diabetes. *Diabetes Obes Metab* 2013;15:189–91.

[67] He R, Zhang D, Lu W, Zheng T, Wan L, Liu F, et al. SLC47A1 gene rs2289669 G>A variants enhance the glucose-lowering effect of metformin via delaying its excretion in Chinese type 2 diabetes patients. *Diabetes Res Clin Pract* 2015;109:57–63.

[68] Liang H, Xu W, Zhou L, Yang W, Weng J. Differential increments of basal glucagon-like-1 peptide concentration among SLC47A1 rs2289669 genotypes were associated with inter-individual variability in glycaemic response to metformin in Chinese people with newly diagnosed Type 2 diabetes. *Diabet Med J Br Diabet Assoc* 2017;34:987–92.

[69] Mousavi S, Kohan L, Yavarian M, Habib A. Pharmacogenetic variation of SLC47A1 gene and metformin response in type2 diabetes patients. *Mol Biol Res Commun* 2017;6:91–4.

[70] Stocker SL, Morrissey KM, Yee SW, Castro RA, Xu L, Dahlin A, et al. The effect of novel promoter variants in MATE1 and MATE2 on the pharmacokinetics and pharmacodynamics of metformin. *Clin Pharmacol Ther* 2013;93:186–94.

[71] Klen J, Goričar K, Janež A, Dolžan V. The role of genetic factors and kidney and liver function in glycemic control in type 2 diabetes patients on long-term metformin and sulphonylurea cotreatment. *Biomed Res Int* 2014;2014:934729.

[72] Tarasova L, Kalnina I, Geldnere K, Bumbure A, Ritenberga R, Nikitina-Zake L, et al. Association of genetic variation in the organic cation transporters OCT1, OCT2 and multidrug and toxin extrusion 1 transporter protein genes with the gastrointestinal side effects and lower BMI in metformin-treated type 2 diabetes patients. *Pharmacogenet Genomics* 2012;22:659–66.

[73] Christensen. The pharmacogenetics of metformin and its impact on plasma metformin steady-state levels and glycosylated hemoglobin A1c: Corrigendum. *Pharmacogenet Genomics* 2015;25:48.

[74] Raj GM, Mathaiyan J, Wyawahare M, Priyadarshini R. Lack of effect of the SLC47A1 and SLC47A2 gene polymorphisms on the glycemic response to metformin in type 2 diabetes mellitus patients. *Drug Metab Pers Ther* 2018;33:175–85.

[75] Phani NM, Vohra M, Kakar A, Adhikari P, Nagri SK, D'Souza SC, et al. Implication of critical pharmacokinetic gene variants on therapeutic response to metformin in Type 2 diabetes. *Pharmacogenomics* 2018;19:905–11.

[76] Cho SK, Chung J-Y. The MATE1 rs2289669 polymorphism affects the renal clearance of metformin following ranitidine treatment. *Int J Clin Pharmacol Ther* 2016;54:253–62.

[77] Lechner C, Ishiguro N, Fukuhara A, Shimizu H, Ohtsu N, Takatani M, et al. Impact of Experimental Conditions on the Evaluation of Interactions between Multidrug and Toxin Extrusion Proteins and Candidate Drugs. *Drug Metab Dispos Biol Fate Chem* 2016;44:1381–9.

[78] Martínez-Guerrero LJ, Morales M, Ekins S, Wright SH. Lack of Influence of Substrate on Ligand Interaction with the Human Multidrug and Toxin Extruder, MATE1. *Mol Pharmacol* 2016;90:254–64.

[79] Chowdhury S, Yung E, Pintilie M, Muaddi H, Chaib S, Yeung M, et al. MATE2 Expression Is Associated with Cancer Cell Response to Metformin. *Plos One* 2016;11:e0165214.

[80] Morrissey KM, Stocker SL, Chen EC, Castro RA, Brett CM, Giacomini KM. The Effect of Nizatidine, a MATE2K Selective Inhibitor, on the Pharmacokinetics and Pharmacodynamics of Metformin in Healthy Volunteers. *Clin Pharmacokinet* 2016;55:495–506.

[81] Oh J, Chung H, Park S-I, Yi SJ, Jang K, Kim AH, et al. Inhibition of the multidrug and toxin extrusion (MATE) transporter by pyrimethamine increases the plasma concentration of metformin but

does not increase antihyperglycaemic activity in humans. *Diabetes Obes Metab* 2016;18:104–8.

[82] Toyama K, Yonezawa A, Tsuda M, Masuda S, Yano I, Terada T, et al. Heterozygous variants of multidrug and toxin extrusions (MATE1 and MATE2-K) have little influence on the disposition of metformin in diabetic patients. *Pharmacogenet Genomics* 2010;20:135–8.

[83] Dujic T, Zhou K, Yee SW, van Leeuwen N, de Keyser CE, Javorský M, et al. Variants in Pharmacokinetic Transporters and Glycemic Response to Metformin: A Metgen Meta-Analysis. *Clin Pharmacol Ther* 2017;101:763–72.

[84] Zhou K, Yee SW, Seiser EL, van Leeuwen N, Tavendale R, Bennett AJ, et al. Variation in the glucose transporter gene SLC2A2 is associated with glycemic response to metformin. *Nat Genet* 2016;48:1055–9.

[85] Rathmann W, Strassburger K, Bongaerts B, Kuss O, Müssig K, Burkart V, et al. A variant of the glucose transporter gene SLC2A2 modifies the glycaemic response to metformin therapy in recently diagnosed type 2 diabetes. *Diabetologia* 2019;62:286–91.

[86] Zhou M, Xia L, Wang J. Metformin transport by a newly cloned proton-stimulated organic cation transporter (plasma membrane monoamine transporter) expressed in human intestine. *Drug Metab Dispos Biol Fate Chem* 2007;35:1956–62.

[87] Han TK, Proctor WR, Costales CL, Cai H, Everett RS, Thakker DR. Four cation-selective transporters contribute to apical uptake and accumulation of metformin in Caco-2 cell monolayers. *J Pharmacol Exp Ther* 2015;352:519–28.

[88] Wang J. The plasma membrane monoamine transporter (PMAT): Structure, function, and role in organic cation disposition. *Clin Pharmacol Ther* 2016;100:489–99.

[89] Arner P, Kulyté A, Batchelor K, Laurencikiene J, Livingston J, Rydén M. Mapping of biguanide transporters in human fat cells and their impact on lipolysis. *Diabetes Obes Metab* 2018;20:2416–25.

[90] Duong JK, Kumar SS, Kirkpatrick CM, Greenup LC, Arora M, Lee TC, et al. Population pharmacokinetics of metformin in healthy subjects and patients with type 2 diabetes mellitus: simulation of doses according to renal function. *Clin Pharmacokinet* 2013;52:373–84.

[91] Rena G, Hardie DG, Pearson ER. The mechanisms of action of metformin. *Diabetologia* 2017;60:1577–85.

[92] He L, Wondisford FE. Metformin action: concentrations matter. *Cell Metab 2015*;21:159–62.

[93] GoDARTS and UKPDS Diabetes Pharmacogenetics Study Group, Wellcome Trust Case Control Consortium 2, Zhou K, Bellenguez C, Spencer CCA, Bennett AJ, et al. Common variants near ATM are associated with glycemic response to metformin in type 2 diabetes. *Nat Genet* 2011;43:117–20.

[94] Woods A, Leiper JM, Carling D. The role of ATM in response to metformin treatment and activation of AMPK. *Nat Genet* 2012;44:360–1.

[95] Yee SW, Chen L, Giacomini KM. The role of ATM in response to metformin treatment and activation of AMPK. *Nat Genet.* 2012;44:359–60.

[96] Florez JC, Jablonski KA, Taylor A, Mather K, Horton E, White NH, et al. The C allele of ATM rs11212617 does not associate with metformin response in the Diabetes Prevention Program. *Diabetes Care* 2012;35:1864–7.

[97] Van Leeuwen N, Nijpels G, Becker ML, Deshmukh H, Zhou K, Stricker BHC, et al. A gene variant near ATM is significantly associated with metformin treatment response in type 2 diabetes: a replication and meta-analysis of five cohorts. *Diabetologia* 2012;55:1971–7.

[98] Tkáč I. Replication of the association of gene variant near ATM and response to metformin. *Pharmacogenomics* 2012;13:1331–2.

[99] Zhou Y, Guo Y, Ye W, Wang Y, Li X, Tian Y, et al. RS11212617 is associated with metformin treatment response in type 2 diabetes in Shanghai local Chinese population. *Int J Clin Pract* 2014;68:1462–6.

[100] Al-Khalili L, Forsgren M, Kannisto K, Zierath JR, Lönnqvist F, Krook A. Enhanced insulin-stimulated glycogen synthesis in response to insulin, metformin or rosiglitazone is associated with increased mRNA expression of GLUT4 and peroxisomal proliferator activator receptor gamma co-activator 1. *Diabetologia* 2005; 48:1173–9.

[101] Dawed AY, Ali A, Zhou K, Pearson ER, Franks PW. Evidence-based prioritisation and enrichment of genes interacting with metformin in type 2 diabetes. *Diabetologia* 2017;60:2231–9.

[102] Martono DP, Heerspink HJL, Hak E, Denig P, Wilffert B. No significant association of type 2 diabetes-related genetic risk scores with glycated haemoglobin levels after initiating metformin or sulphonylurea derivatives. *Diabetes Obes Metab* 2019.

[103] Heo CU, Choi C-I. Current Progress in Pharmacogenetics of Second-Line Antidiabetic Medications: Towards Precision Medicine for Type 2 Diabetes. *J Clin Med* 2019;8(3).

[104] Ordelheide A-M, Hrabě de Angelis M, Häring H-U, Staiger H. Pharmacogenetics of oral antidiabetic therapy. *Pharmacogenomics* 2018;19:577–87.

[105] Zhou K, Pedersen HK, Dawed AY, Pearson ER. Pharmacogenomics in diabetes mellitus: insights into drug action and drug discovery. *Nat Rev Endocrinol* 2016;12:337–46.

[106] Pearson ER. Pharmacogenetics and target identification in diabetes. *Curr Opin Genet Dev* 2018;50:68–73.

[107] Koivula RW, Forgie IM, Kurbasic A, Viñuela A, Heggie A, Giordano GN, et al. Discovery of biomarkers for glycaemic deterioration before and after the onset of type 2 diabetes: descriptive characteristics of the epidemiological studies within the IMI DIRECT Consortium. *Diabetologia* 2019;62:1601–15.

In: The Pharmacological Guide to Metformin ISBN: 978-1-53616-634-7
Editor: Natalie Lajoie

Chapter 2

PHYSICOCHEMICAL AND PHARMACOKINETIC PROPERTIES OF METFORMIN HYDROCHLORIDE

Selma Sahin*, PhD and Tugba Gulsun, PhD
Department of Pharmaceutical Technology, Faculty of Pharmacy,
Hacettepe University, Ankara, Turkey

ABSTRACT

Metformin hydrochloride is a widely used biguanide derivative antidiabetic agent for the treatment type 2 diabetes mellitus. Metformin does not improve insulin secretion and produce normally hypoglycemia. Metformin decreases blood glucose levels by increasing skeletal muscle uptake of glucose, decreasing gluconeogenesis and absorption of glucose. Metformin has been found to be effective in the treatment of especially prostate, colon and breast cancer. Also, it is reported that metformin treatment decreases the risk of developing type 2 diabetes mellitus in women with polycystic ovarian syndrome. Metformin has a cationic charge at physiological pH and has very limited passive diffusion across cell membranes. According to Biopharmaceutics Classification System,

* Corresponding Author's E-mail: sahin.selma@gmail.com.

metformin hydrochloride is a class 3 drug with high solubility and low permeability properties. Its absorption is slow and incomplete, and thus has low bioavailability (50-60%) following oral administration. Metformin is a substrate of organic cation transporters 1, 2 and 3 (OCT1-3) and its oral absorption, hepatic uptake and renal excretion occur mainly via these transporters. Membrane monoamine transporter (PMAT) expressed in human intestine may play a role in the intestinal absorption of metformin. Additionally, metformin is a good substrate for human multidrug and toxin extrusion 1 (MATE1) and MATE2-K. Metformin displays flip–flop kinetics (in human $t_{1/2,po}/t_{1/2,iv}$ =4.1) after oral administration. Its binding to human plasma proteins is negligible after absorption. Metformin accumulates in the kidney, salivary glands and walls of the small intestine, and it has a large apparent volume of distribution. The average elimination half-life in plasma is about 6 hours. Metformin is not metabolized, and it is eliminated from the body by tubular secretion and excreted unchanged in the urine. Physicochemical, pharmacological, and pharmacokinetic properties of metformin will be reviewed this chapter.

Keywords: metformin hydrochloride, diabetes mellitus, physicochemical characteristics, pharmacokinetic parameters, pharmacologic properties

INTRODUCTION

Diabetes mellitus (commonly known as diabetes) is a major health risk in many countries, and the incidence rates are increasing [1] because of aging, physical inactivity and obesity prevalence.The World Health Organization (WHO) predicts that future diabetes prevalence will be increased approximately 1.5-fold (from 425 million to 629 million adults) in 28 years (2017-2045) [2]. Diabetes is associated with many complications such as coronary artery disease, cerebrovascular diseases, peripheral vascular disease, stroke, retinopathy, nephropathy, neuropathy, diabetic foot, therefore, the money spent for the diagnosis, treatment and complications of diabetes is an important burden for the world economy [3].

Diabetes mellitus is a metabolic disorder that damages insulin secretion and insulin action or both. Insulin deficiency also leads to chronic hyperglycaemia as a result of carbohydrate, fat, and protein metabolism disorders [4]. It is characterized by hyperglycemia (high blood sugar) in the context of insulin resistance and relative lack of insulin [5, 6]. The first commonly accepted classification of diabetes mellitus was announced by WHO in 1980, and revised in 1985. Today, American Diabetes Association® has made a new classification of diabetes based on its etiology. According to this classification, four types of diabetes mellitus are described namely type 1, gestational, other specific types and type 2 diabetes. New classification is based on pathogenesis, not insulin dependence [7].

Type 1 diabetes (also known as insulin-dependent, juvenile or childhood-onset) is caused by the lack of insulin and patients use insulin for treatment. The cause of type 1 diabetes is unknown and cannot be avoided by the available information. Gestational diabetes occurs during pregnancy, is characterized by high glucose levels. Genetic defects of beta-cell function, genetic defects in insulin action, diseases of the exocrine pancreas, endocrinopathies, drug or chemical induced, infections, uncommon forms of immune-mediated diabetes, other genetic syndromes sometimes associated with diabetes are subgroups of other types of diabetes [8]. Type 2 diabetes mellitus (also known as non-insulin-dependent, or adult-onset) is a disease of insulin resistance by cells. Although until quite recently, this type of diabetes was seen only in adults, it is increasingly common in children as well. Type 2 diabetes is the most common type of diabetes. The number of people with type 2 diabetes is increasing as a result of aging, physical inactivity, and obesity prevalence [9]. Patients use oral antidiabetic agents for treatment of type 2 diabetes. Several groups of oral antidiabetic drugs such as biguanides (active compound of French Lilac; Galega officinalis), sulfonylureas, thiazolidinediones, alpha-glucosidase inhibitors are used for the treatment of type 2 diabetes. Of these, biguanide -derivative metformin is recommended as the first-line medication for treatment of type-2 diabetes by leading authorities such as the International Diabetes Federation, the American Diabetes Association®. Metformin HCl is given orally in tablets containing 500 mg, 850 mg or 1000 mg of metformin HCl for treatment type 2 diabetes in adults and children. The initial dose is 500 mg or 850 mg twice a day with meals. Maintenance dose is 500 and 850 mg 2-3 times a day with meals. The maximum dose is 2550 mg [10]. Metformin extended release tablets (500 mg, 750, and 1000 mg metformin HCl) are commercially available for adult use only. Safety and efficacy of these extended release tablets in children have not been studied yet. Fixed dose combination of metformin with other antidiabetic agents (rosiglitazone, glyburide, sitagliptin, linagliptin, alogliptin, saxagliptin, glipizide and repaglinide) are also commercially available [11]. The aim of this chapter is to discuss physicochemical, pharmacological, and pharmacokinetics properties of metformin [12].

PHYSICOCHEMICAL PROPERTIES

Metformin was synthesized in 1920s, and it was approved by the Food and Drug Administration (FDA) in 1995 for the treatment of type 2 diabetes [13].

Metformin is N,N-dimethylimidodicarbonimidic diamide ($C_4H_{11}N_5$, Molecular mass: 129.17 g/mol; Metformin is a small amphoteric molecule with pKa values of 2.8 and 11.5. It has a logD of -3.37 at pH 4.0 [14]. Metformin (used as the HCl salt) is a white crystalline powder with a molecular weight of 165.63 g/mol (Figure 1) [15]. It is freely soluble in water; slightly soluble in alcohol; practically insoluble in acetone and in dichloromethane [16, 17]. Partition coefficient of metformin HCl is 0.0622±0.0021 in n-octanol:water, and 0.0541±0.0016 in n-octanol: phosphate buffered saline (pH 7.4). Its melting point is between 222-226 °C [17, 18]. The λ_{max} for ultraviolet detection of metformin in methanol is 236 nm [19]. In human plasma, metformin is stable for 48 hours at room temperature, and 2 years at -80°C. Metformin HCl is also stable at UV light, heat, acids, bases, peroxidation and oxidative stress conditions [20, 21]. Metformin HCl should be stored at room temperature below 25°C.

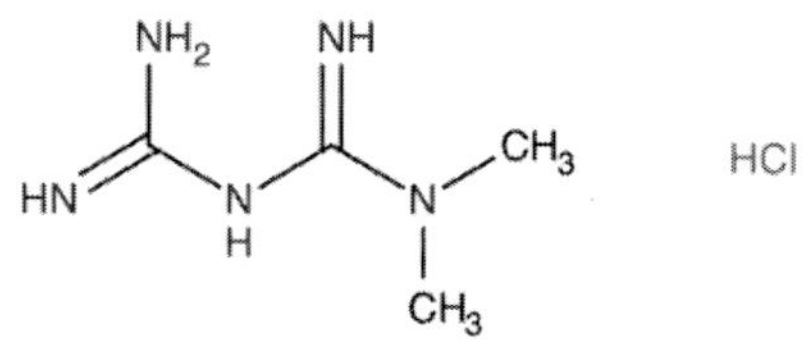

Figure 1. Chemical structure of metformin HCl [20, 22].

Metformin HCl has no known polymorphs, hydrates or solvates formed at room temperature. Metformin HCl is highly water-soluble throughout the physiological pH range (> 100 mg/mL) and has very low lipid solubility. Metformin has a cationic charge at physiological pH (cationic percentage at physiologic pH > 99.9%) [23-25]. Therefore, its passive diffusion across cell membranes is very limited, and it is transported by an active transport mechanism. It has a low permeability across the cell membranes [26]. Various methods have been used for determination of permeability values of metformin in the literature. Its permeability is in the range of 3.7×10^{-7} - 19×10^{-6} cm/s across Caco-2 monolayers [27-30]. Similar apparent permeability coefficients were also reported using everted gut sac technique (7.37×10^{-7} cm/s) and Sartorius SM 16750 Absorption Simulator (6.13×10^{-7} cm/s) [31]. On the other hand, higher intestinal permeability values were reported for metformin (50 µg/mL) using single-pass intestinal perfusion studies in rats (4.51×10^{-5} cm/s for duodenum, 3.26×10^{-5} cm/s for jejunum, and 2.96×10^{-5} cm/s for ileum). It was also reported that permeability of metformin across duodenum was influenced by metformin concentration used. With an increase in metformin concentration from 10 µg/mL to 200 µg/mL, its permeability decreased significantly from 4.71×10^{-5} cm/s to 2.70×10^{-5} cm/s [32]. Based on its aqueous solubility and permeability, metformin is classified as Class 3 compound (high solubility, low permeability/poor metabolism) according to the Biopharmaceutics Classification System (BCS) and Biopharmaceutics Drug Disposition Classification System (BDDCS). Therefore, following oral administration, permeability is the rate the limiting step for its absorption, and hence bioavailability [33-36].

Also, it is classified as a Class 2 compound according to the Salivary Excretion Classification System (SECS) [37].

IDENTIFICATION AND QUANTIFICATION METHODS

Metformin HCl in pure form, in pharmaceutical formulations, in the combination of pharmaceutical products can be determined via spectrophotometric (UV-Visible Spectrophotometer) and colorimetric methods [38, 39]. Numerous gas chromatography [40], gas chromatography-mass spectrometry (GC-MS) [41, 42], high-pressure liquid chromatography (HPLC) [43-46] capillary electrophoresis [47, 48]. LC–MS/MS [49, 50] LC-MS/MS-ESI [51, 52], hydrophilic interaction liquid chromatography (HILIC) [53], high performance thin layer chromatography-densitometry (HPTLC) [54], ion selective electrochemical method (ISE) [55], tandem mass spectrometry [56, 57] or micellar chromatography [58], spectrofluorometric methods [59, 60] volumetric methods [61] are used for determination of metformin HCl in human and rat plasma and urine. The methods are most often based on traditional reverse phase chromatography and cation exchange or ion pairing. Additionally, liquid scintillation counting is also used for determination of ^{14}C labeled metformin HCl in humans [62]. Among these methods, reverse phase HPLC method is the most commonly used quantitative method for determination of metformin HCl in rabbit, rat or human plasma and urine, human serum, pharmaceutical formulations (e.g., tablets, nanoparticles) or in vitro studies because of its simplicity, rapidity, high specificity, sensitivity, reproducibility, and accuracy. All these developed HPLC methods differ as to mobile phase, columns and detection methods. Additionally, quite a number of methods are available in the literature for simultaneous analysis of metformin with other drugs such as linagliptin, glimepride, vildagliptin, saxagliptin, sitagliptin, pioglitazone, diltiazem, rosiglitazone, cimetidine, famotidine, ranitidine, glyburide [63-71].

PHARMACOKINETIC PROPERTIES

Absorption and Bioavailability

Metformin HCl is slowly absorbed from the upper part of the small intestine. Its gastrointestinal absorption is completed within 6 hours of ingestion [72, 73]. Permeability is the rate-limiting step for metformin absorption. Its absolute bioavailability is approximately 50-60% following oral administration of 500 mg single dose, and shows intra-intersubject variability [13, 74]. Various drug delivery systems, such as bioadhesive and gastroretentive drug delivery systems can be used to enhance its incomplete absorption [36, 74-77]. Food delays and reduces absorption and bioavailability of metformin (approximately 40% lower C_{max}, 25% lower AUC, and 35 minutes longer t_{max}) [23, 78-80]. It is reported that there is a lack of dose proportionality with increasing doses due to decreased absorption rather than alteration in elimination [81]. Metformin shows flip-flop pharmacokinetic (in human $t_{1/2,po}/t_{1/2,iv}$ =4.1) because its rate of absorption is slower than elimination [82-84]. After single oral doses (500, 850 and 1500 mg), peak plasma concentration (C_{max}) is reached within 2 to 3.3 hours [85, 86]. Following oral administration of immediate release-metformin tablet (1000 mg) twice daily, the higher C_{max} values were reported in type 2 diabetic patients with good renal function (2.09 mg/L) compared to healthy subjects (1.32 mg/L) [23]. C_{max} is prolonged (4-8 hours) for extended release tablets [82]. Evaluation of pharmacokinetic properties of metformin-extended release tablets (750 mg) under fasting and fed states (after high-fat breakfast) revealed that bioavailability and time to reach C_{max} increased in the fed state in comparison to the fasted state, however, C_{max} (794 ng/mL versus 832 ng/mL) and half-life ($t_{1/2}$: 3.66 h versus 3.80 h) values were very similar for both conditions. On the other hand, C_{max}, t_{max}, and AUC values obtained for immediate release tablets (1000 mg, under fasting conditions) were significantly different from the corresponding values obtained for the extended release formulation [37]. Pharmacokinetic parameters of metformin in human gathered from the literature are given in Table 1.

Table 1. Pharmacokinetic parameters of metformin in human

Ref	Administration		C_{max} (ng/mL)	t_{max} (h)	AUC (ng.h/mL)	$t_{1/2}$ (h)	CL_R (mL/min)	CL/F (mL/min)	F (%)	Vd/F (L)
[91]	Metformin Solutions 850 mg (n=24; healthy volunteers)		1500 (200)	2.5 (1.0)	-	4.7 (2.8)	615.00 (141.66)	-	-	-
[92]	Metformin Solutions 1000 mg (n=6; healthy volunteers)		1880 (110)	3.5 (0.3)	-	2.14 (0.07)	596.66 (35.00)	-	52 (3)	-
[93]	Metformin: Glyburide (250;1.25 mg) Tablet (n=21)		640.5 (100.5)	3.3 (0.7)	AUC_{0-12} 4153 (725.1)	-	-	16.66 (9.0)	-	-
[62]	Metformin Tablets 500 mg (n=5; healthy volunteers)		1550 (2400)	1.9 (0.4)	-	2.63(0.18)	443.33(23.33)	-	51.6 (5.2)	-
[33]	Metformin Tablets 500 mg (n=4; healthy volunteers)		1020 (170)	2.2 (0.2)	-	5.43 (0.73)	525.00 (61.66)	1318.33 (196.66)	55 (8)	-
[94]	Metformin Tablets 500 mg (n=22)		1420 (70)	2.59 (0.18)	-	1.54	-	1102 (76.6)	-	26.63
[86]	Metformin 500 mg (n=10)		2000 (400)	-	13900 (390)	4.1 (0.3)	293.33 (131.66)	586.66 (96.66)	-	-
[87]	Metformin Tablets 500 mg (n=24)	Test	1036.08 (192.98)	2.88 (0.85)	$AUC_{0-\infty}$ 7293.19 (1573.26)	3.40 (0.83)	-	-	-	-
		Reference	1016.08 (224.44)	2.95 (0.81)	$AUC_{0-\infty}$ 6996.56 (1427.62)	3.27 (0.74)	-	-	-	-
[95]	Metformin Tablets 850 mg (n=24; healthy volunteers)		1500 (130)	3.3 (0.3)	-	2.39 (0.08)	-	1850 (211.66)	-	-
[96]	Metformin Tablets 850 mg (n=9; healthy volunteers)		1810 (200)	2.62 (0.23)	$AUC_{0-\infty}$ 12230 (1100)	4.6 (0.7)	525.0 (57.2)	968.3(91.7)	-	624.7(159.0)

Ref	Administration	C_{max} (ng/mL)	t_{max} (h)	AUC (ng.h/mL)	$t_{1/2}$ (h)	CL_R (mL/min)	CL/F (mL/min)	F (%)	Vd/F (L)
[96]	Metformin Tablets 850 mg (n=9; diabetic volunteers)	1510 (150)	3.28 (0.38)	$AUC_{0-\infty}$ 11650 (1270)	7.2 (1.9)	481.7 (46.1)	1033.3 (100.0)	-	366.9 (41.6)
[85]	Metformin Tablets 850 mg (n=6)	1700 (100)	3.0 (0.3)	8900 (400)	2.7 (0.2)	-	-	-	-
[97]	Metformin Tablets 850 mg (n=6; healthy young patients)	1390 (330)	2.67 (0.82)	$AUC_{0-\infty}$ 9980 (2120)	6.9 (4.2)	636 (84)	1155 (273)	-	723 (588)
[97]	Metformin Tablets 850 mg (n=12; healthy elderly patients)	2450 (700)	2.71 (1.05)	$AUC_{0-\infty}$ 15990 (3800)	11.3 (7.3)	412 (98)	728 (180)	-	727 (501)
Ref	Administration	C_{max} (ng/mL)	t_{max} (h)	AUC (ng.h/mL)	$t_{1/2}$ (h)	CL_R (mL/min)	CL/F (mL/min)	F (%)	Vd/F (L)
[97]	Metformin Tablets 850 mg (n=3; healthy middle-aged patients)	1640 (500)	3.00 (1.00)	$AUC_{0-\infty}$ 11220 (3190)	11.2 (5.2)	394.7 (83.8)	1031.7 (251.8)	-	43.7 (324.1)
[98]	Metformin Tablets 1000 mg (n=23)	1420 (320)	6.3 (1.4)	$AUC_{0-\infty}$ 11940 (2710) AUC_{0-72} 11900 (2760)	5.00	-	-	-	-
[99]	Metformin Extended Release Tablets 1000 mg (n=16)	1080 (259)	7 (4-8)	AUC_{0-24} 12587 (3164)	4.5 (0.8)	-	-	-	-
[98]	Metformin Tablets 1500 mg (n=23)	1780 (370)	6.7 (1.6)	$AUC_{0-\infty}$ 16700 (4150) AUC_{0-72} 16680 (4140)	5.60	-	-	-	-

Table 1. (Continued)

Ref	Administration	C_{max} (ng/mL)	t_{max} (h)	AUC (ng.h/mL)	$t_{1/2}$ (h)	CL_R (mL/min)	CL/F (mL/min)	F (%)	Vd/F (L)
[33]	Metformin Tablets 1500 mg (n=4; healthy volunteers)	3100 (460)	1.5 (0.2)	-	5.98 (0.75)	518.33 (7.66)	1551.66 (173.33)	50 (10)	-
[96]	Metformin Tablets 1700 mg (n=9; healthy volunteers)	1300 (90)	2.77 (0.48)	9100 (650)	6.3 (0.8)	455.0 (38.9)	1270.0 (99.4)	-	999.2 (214.8)
[96]	Metformin Tablets 1700 mg (n=9; diabetic volunteers)	1305 (90)	3.12 (0.48)	9450 (690)	9.5 (1.9)	475.0 (46.1)	1210.0 (71.1)	-	657.3 (67.5)
[98]	Metformin Tablets 2000 mg (n=23)	2110 (520)	7.8 (2.0)	$AUC_{0-\infty}$ 20810 (3870) AUC_{0-72} 20650 (3820)	7.40	-	-	-	-
[98]	Metformin Tablets 2500 mg (n=23)	2480 (530)	7.2 (2.1)	$AUC_{0-\infty}$ 24260 (4100) AUC_{0-72} 24180 (3970)	7.50	-	-	-	-

The data are given as the mean (standard deviation); CL_R is renal clearance; CL is total clearance, F is bioavailability, V_d is volume of distribution.

A bioequivalence study was performed on 24 healthy volunteers for two metformin-containing formulations (a reference and generic drug) after a single dose of 500 mg metformin tablets. The 90% confidence interval values estimated for C_{max} and AUC ratios were within the 80.00–125.00% interval indicating that generic tablet was bioequivalent to the reference tablet [87]. Similarly, two separate bioequivalence trials performed on healthy volunteers using one test and two reference products containing 500 mg metformin. Based on the 90% confidence interval values calculated for C_{max} and AUC ratios, both reference drugs were found bioequivalent (0.80 - 1.25 for both parameters) [88].

According to the EMA and FDA guidelines, immediate release dosage forms of BCS Class 3 compounds are subject to a BCS-based biowaiver [89, 90]. The recommended criteria for the BCS-based biowaiver are: high solubility and limited absorption for drug substance; very rapid in vitro dissolution rate (>85% in 15 min) for test and reference drug products; and no change in excipients that might affect bioavailability. Metformin HCl is a BCS Class 3 compound, therefore, metformin HCl containing immediate release tablets can be a candidate for BCS-based biowaiver given that all recommended criteria are met.

DISTRIBUTION

Plasma protein binding of metformin is negligible after absorption. Its mean apparent volume of distribution after single oral doses (850 mg) is 654±358 L [100]. The plasma elimination half-life of metformin that exhibits multicompartmental characteristics, is about 2–6 h after oral doses. The steady-state plasma concentrations (54 to 4133 ng/mL) are reached within 24-48 hours [101].

Ionization of metformin in the blood is less than 0.01%, however, it is widely distributed in most tissues such as kidneys, adrenal glands, pancreas, liver, lung, muscle, spleen, and it has a large apparent volume of distribution (Table 1). It does partition into erythrocytes as a function of time [23].

Metformin is excreted into breast milk [102] but its concentration is low in breast milk, and the mean infant exposure to the drug is only 0.28% of the weight-normalized maternal dose indicating that metformin use in breastfeeding mothers is safe [14]. Metformin readily crosses the placenta but it is not reported to be teratogenic [103]. It was reported that metformin does not carry a risk for the fetus during the first trimester [104].

METABOLIZATION AND ELIMINATION

Metformin does not undergo hepatic metabolism (no identified metabolites in human) or biliary excretion [23, 62]. Metformin is renally eliminated by both filtration and active tubular secretion and excreted unchanged in the urine [74, 105]. Renal clearance which is greater than creatinine clearance is taken as an indication of tubular secretion is the major route of elimination for metformin. Metformin is a water soluble drug, therefore, it does not undergo passive resorption in the kidney. Following i.v. administration of ^{14}C-metformin HCl, its elimination half-life from human plasma was 1.7±0.1 h, and it was completely eliminated by renal route with a renal clearance of 454 ±47 mL/min [62]. In another study, after i.v. administration of 250- 1000 mg metformin, its terminal half-life was 1.5-4.5 hours, and 79-100% of metformin was excreted renally in human [106]. Following oral administration, approximately 90% of the absorbed drug is eliminated via the renal route within the first 24 hours, with a plasma elimination half-life of approximately 6.2 hours. In blood, the elimination half-life is approximately 17.6 hours, suggesting that the erythrocyte may be a compartment for distribution [107]. About 20-30% of an oral dose of metformin HCl is recovered in the feces [33]. Metformin is a substrate for several transporters in the kidney [23]. In healthy volunteers, evaluation of metformin renal clearance in the presence a MATE inhibitor pyrimethamine revealed that pyrimethamine significantly reduced the renal clearance of metformin by 35% [108].

On the other hand, metformin is metabolized by liver mainly via CYP2C11, 2D1, 3A1/2 in rats [109]. In cats, it was also renally eliminated (renal clearance 0.13±0.03 L/h/kg) with a negligible nonrenal clearance (0.02+/-0.02 L/kg) [110].

METFORMIN AND DRUG TRANSPORTERS

Metformin is a substrate of organic cation transporters 1-3 (OCT1, OCT2, OCT3) [111] expressed predominantly in the liver and kidneys. OCT1 (and possibly OCT3) contributes to hepatic uptake of metformin [112, 113]. OCT2 is expressed on the basolateral membrane of kidney tubular cells, and is responsible for 80% of the total metformin clearance [114]. OCT3 is localized on the brush border membrane of human small intestinal epithelial cells, and contributes to intestinal absorption of metformin [115]. All these OCTs (OCT1, OCT2 and OCT3) may have a significant effect on the pharmacokinetics of metformin [116-119]. Following i.v. infusion of metformin to wild type and Oct 1 (-/-) knockout mice at a dose of 150 mg/h/kg for 210 min, hepatic concentration of metformin in Oct1(-/-) was less than those of wild type mice (49.3±10.4 μg/g vs 417±178 μg/g), whereas, concentration of metformin in the skeletal muscle was comparable between Oct1(-/-) (107±17 μg/g) and wild-type mice (153±18 μg/g) [120]. In another study, Oct1/Oct2 double-knockout and wild-type FVB male mice were used to investigate effect of transporters on metformin pharmacokinetics. After i.v. administration of metformin (5 mg/kg) to the mice, all plasma concentrations were higher in Oct1/2-knockout mice. Systemic exposure was on average 4.4-fold higher in Oct1/2-knockout mice. Metformin clearance in wild type mice (81.7 mL/minxkg) was higher than Oct1/2-knockout mice (18 mL/minxkg). Volume of distribution was 3.5-fold higher in wild type mice [121].

Plasma membrane monoamine transporter (PMAT) that is localized on the luminal side of enterocytes, may be the major transporter responsible for gastrointestinal uptake of metformin HCl into cells [122]. PMAT is also expressed on the apical membrane of renal epithelial cells may be implicated in the renal reabsorption of metformin [122, 123].

Furthermore, metformin is also a good substrate for human multidrug and toxin extrusion 1 (MATE1) and MATE2-K. These transporters are localized on the apical membrane of the renal proximal tubule cells, and contribute to the renal excretion of metformin from the tubule cell to the lumen, and transport of metformin into bile canaliculus. There are many studies in literature supporting the role of MATE1 in renal elimination of metformin [108, 112, 124, 125].

In addition, metformin is also a substrate of organic cation transporter novel, type 1 (OCTN1) and thiamine transporter 2 (THTR-2) which may affect its absorption from intestine and reabsorption from kidney [126]. It was demonstrated that OCTN1 mediates the uptake of metformin in human embryonic kidney 293 (HEK293) cells transfected with the mouse [127].

As a result, all these transporters may have an effect on oral absorption, hepatic uptake and renal excretion of metformin.

Pharmacological Properties

Mechanism of Action

Although metformin has been used for the treatment of type 2 diabetes since 1950's, its mechanism of action has not been fully understood yet. Its mechanisms of action are different from other classes of oral antidiabetic agents. Biguanide derivate of metformin HCl does not produce hypoglycemia in either type 2 diabetic patients or healthy subjects, and does not cause hyperinsulinemia. Therefore, metformin is classified as an antihyperglycemic agent not a hypoglycemic agent [83].

Various mechanisms of action have been proposed for the pharmacological effect of metformin in diabetes. The first proposed mechanism is that metformin HCl accumulates in the intestinal wall, and thus decreases gastrointestinal absorption of glucose [128]. The other mechanism is that metformin decreases glucose production in the liver by inhibiting gluconeogenesis and glycogenolysis. There are numerous studies in the literature confirming that metformin inhibits hepatic glucose production in diabetic patients [129-131]. In addition, metformin increases glucose uptake and the use of glucose in the intestines, skeletal muscle and adipose tissue [132]. Metformin increases insulin sensitivity by increasing peripheral glucose uptake and use in the muscle.

Recently, metformin has been reported to increase glycaemia by affecting the liver via AMPK activation. Metformin inhibits the mitochondrial respiratory chain in the liver. Then, AMPK activated and improve insulin sensitivity by affecting fat metabolism. cAMP eventually decreases and therefore, the expression of gluconeogenic enzymes decrease [133].

Recent studies demonstrated that metformin has an influence on the modulation of multiple components of the incretin axis. Plasma level of glucagon-like peptide 1 is increased, and incretin receptor gene expression in islets cells via peroxisome proliferator-activated receptor PPAR-α is stimulated by metformin. Furthermore, metformin increases insulin secretion by stimulating the function of GLP-1 on β-cells [134-136].

Uses and Administration

Type 2 diabetes which is associated with insulin resistance in several organs (e.g., muscle, liver and adipose tissue), effects multiple organ systems, and decline in β-cell function due to the body's ineffective use of insulin. Microvascular and/or macrovascular complications are usually determined during the diagnosis of type 2 diabetes [137]. Its prevalence is increasing as a result of excess body weight, physical inactivity and unhealthy diet [137-139]. Therefore, in the treatment of type 2 diabetic patients, healthy diet, regular physical activity and weight loss are usually recommended in addition to pharmacological agents. Metformin HCl is a widely used pharmacological agent indicated for the treatment of type 2 diabetic adult and children patients especially type 2 diabetic obese people whose disease have not been cured after dietary treatment and exercise [140]. It is used either alone or with other antidiabetic agents (sulfonylureas, alpha-glucosidase inhibitors) or in combination with insulin. A limited number of studies have considered the addition of metformin to insulin therapy in type 1 diabetics [141, 142].

Immediate release (500, 625, 750, 850, 1000 mg) and extended release dosage forms of metformin HCl (500, 750, 1000 mg) are commercially available for the treatment of type 2 diabetes. Dosage regimens for adults (18-79 years), children (0-9 years and 10-17 years) are summarized in Table 2.

Table 2. Recommended dosage regimens of metformin for adults and children

	Children (0-9 years)	**Children (10-17 years)**	**Adults**
Immediate release	This medication has not been studied in children younger than 10 years of age.	Typical starting dosage: 500 mg taken twice per day.	Typical starting dosage: 500 mg, twice per day, or 850 mg, once per day.
		Dosage changes: increasing dosage by 500 mg every week in divided doses.	Dosage changes: increase dosage by 500 mg weekly or 850 mg every 2 weeks, up to a total of 2550 mg taken per day in divided doses.
		Maximum dosage: 2000 mg per day.	Maximum dosage: 2550 mg per day.
Extended release	This medication has not been studied in children younger than 10 years of age.	This medication has not been studied in children younger than 18 years of age.	Typical starting dosage: 500 mg taken once per day
			Dosage changes: increase dosage by 500 mg every week.
			Maximum dosage: 2000 mg per day.

Polycystic ovary syndrome (PCOS), a common endocrine disorder that effects the reproductive-aged women, is caused by hyperandrogenism, insulin resistance, non-alcoholic fatty liver disease and premature puberty [143, 144]. A strong correlation was reported between hyperandrogenism and hyperinsulinemia. Metformin may regulate the menstrual cycles and increase fertility by increasing insulin sensitivity, reducing insulin level and improving androgen concentrations in women with insulin resistance and dysglicemia [145]. Clinical data indicate that metformin is an effective first-line treatment for restoring ovulatory menstrual cycles in oligoamenorrheic PCOS patients, improving insulin sensitivity and lipoprotein pattern [146].

Some studies have implied that metformin HCl could have therapeutic or renoprotective effects against nephrotoxicity [147, 148]. This effect of metformin is very important in preventing diabetic nephropathy. However, the exact mechanism action of metformin HCl against nephrotoxicity is still unknown.

The anticancer properties of metformin are related to its direct insulin-independent and indirect insulin-dependent actions. Its effects on cancer cell are associated with the activation of AMPK, mitogen-activated protein kinase (MAPK), Src tyrosine kinase, the inhibition of mammalian target of rapamycin signaling, protein synthesis and the reduction of epidermal growth factor receptor (EGFR) [149]. Additionally, metformin causes the accumulation of reactive oxygen species that leads to cell death. It was demonstrated that combination of dichloroacetate with metformin synergistically induce caspase-dependent apoptosis involving oxidative damage [150]. It was shown by the recent studies that metformin might be used in the treatment of hyperinsulinemia associated breast and colon cancers [151]. It was reported that daily use of 500 mg metformin inhibits the breast cancer cells [152]. However, high glucose levels promote the growth and aggression of breast cancer cells, and effect of metformin on breast cancer is reduced by the higher glucose levels [153]. In addition, incidence of colorectal cancer is reduced by the long-term use of metformin in diabetes [154]. It was suggested that co-administration of metformin and chemotherapeutic agents (e.g., oxaliplatin) could be used as an effective treatment regimen for recurring colorectal cancer [155]. Also, its inhibitory effect on the proliferation of prostate, breast, endometrial, ovarian, colon and glioma cells have been demonstrated in the cell culture studies [156, 157]. Metformin is reported to reduce cancer risk and cancer-related mortality in diabetic patient when compared to the diabetic patients not treated with metformin [152, 156].

Several studies have shown that the incidence of cancer in diabetic patients treated with metformin is low compared to patients treated with other antidiabetic drugs [158, 159]. Although studies on this subject are increasing rapidly, there is not clinical use of metformin for cancer yet.

ADVERSE EFFECTS

Metformin is generally well tolerated. However, the most common adverse effects related with metformin are gastrointestinal irritation such as nausea, vomiting, metallic taste in the mouth, diarrhea, epigastric pain or loss of appetite [160]. These problems can be overcome by taking tablets with food. Metformin rarely cause decrease in absorption of vitamin B_{12} and folate in chronic treatment. Although hypoglycemia is rare when metformin is given alone, it can happen when metformin is taken with other drugs (e.g., sulfonylureas and insulin) to lower blood sugar. Metformin administration can cause ovulation in patients with polycystic ovary syndrome. In addition, skin problems, liver problems, abnormal laboratory test results may be occurred very rarely with metformin HCl [82]. Lactic acidosis is a rare but serious side effect of this drug. It can be fatal and requires treatment in the hospital. Metformin HCl is contraindicated in patients with a known renal disease or renal dysfunction, hepatic disease, hypersensitivity to metformin HCl, acute or chronic metabolic acidosis.

PRECAUTIONS

As the major route of elimination for metformin is renal excretion, impairment in renal function can lead to metformin accumulation and lactic acidosis [161]. Therefore, serum creatinine levels of especially elderly patients should be controlled. If the serum creatinine levels are below the required limits, metformin should not be used. If the patients plan to have an injection of a iodinated contrast materials, metformin HCl should not be taken due to acute alteration of renal function [162]. Metformin should be discontinued when hypoxemia related with cardiovascular collapse, acute congestive heart failure, and acute myocardial infarction is seen in patients. Because, all these are associated with lactic acidosis, and may lead to acute renal failure. Taking metformin should be stopped during surgical procedures until renal function is normal. Patients with the hepatic failure should use metformin carefully. Long-term use of metformin may cause a lack of vitamin B_{12} [163], therefore, hematologic tests should be performed in patients at regular intervals. When an abnormal value is determined, necessary precaution should be taken. Fever, trauma, infection, surgery may cause a temporary loss of glycemic control. In this case, metformin HCl treatment has to be interrupted. Before metformin treatment, patients should be checked for serum electrolytes and ketones, blood glucose, blood pH, lactate, pyruvate levels [164]. Metformin HCl is pregnancy category B according to FDA [165]. Although metformin was not teratogenic in rats and rabbits at doses up to 600 mg/kg/day, metformin should not be used in pregnancy unless it is clearly needed. Safety and efficacy of metformin were demonstrated in pediatric patients aged 10 to 16 years [166].

DRUG INTERACTIONS

Metformin does not undergo hepatic metabolism, therefore, there is no metabolic interaction with metformin. On the other hand, OCT transporters are important for metformin interaction with other drugs [167]. It was reported that C_{max} and AUC values of metformin increased and its renal clearence decreased when metformin adminestered with an OCT substrate vandetanib [168]. Additionaly, verapamil strongly inhibits the hepatic uptake of metformin by OCT1, and hence glucose-lowering effect of metformin is decreased [169].

Metformin has drug-drug interactions with various drugs including glyburide, furosemide, nifedipine, cationic drugs such as amiloride, digoxin, morphine, procainamide, quinidine, quinine, ranitidine, triamterene, trimethoprim, and vancomycin [170, 171]. Also, some drugs (thiazides and other diuretics, corticosteroids, phenothiazines, thyroid products, estrogens, oral contraceptives, phenytoin, nicotinic acid, sympathomimetics, calcium channel blocking drugs, isoniazid) can cause hyperglycemia when used with metformin [100]. In addition, it is recommended to be careful when administered with alcohol because alcohol affects lactate metabolism of metformin HCl [172]. Metformin does not alter the bioavailability of propranolol and ibuprofen significantly. Some drug interactions with metformin are summarized in Table 3. Metformin is negligibly bound to plasma proteins; therefore, it does not interact with highly protein bound drugs such as salicylates, sulfonamides, chloramphenicol, and probenecid.

Table 3. Metformin interactions with other drugs [100]

Drugs	Interactions
Glyburide	May cause a decrease in AUC and C_{max} of glyburide but no change in metformin pharmacokinetics
Furosemide	Coadministration with furosemide affects C_{max} (about 22% increase) and AUC (about 15% increase) of metformin but no change in renal clearance of metformin.
Nifedipine	May cause 20% increase in C_{max} and, 9% increase in AUC of metformin, and an increase of the amount of metformin in the urine.
Cationic drugs (e.g., amiloride, digoxin, morphine, procainamide, quinidine, quinine, ranitidine, triamterene, trimethoprim, or vancomycin)	Eliminated by proximal renal tubular secretion system and may influence the pharmacokinetics of metformin.
Other (e.g., thiazides and other diuretics, corticosteroids, phenothiazines, thyroid products, estrogens, oral contraceptives, phenytoin, nicotinic acid, sympathomimetics, calcium channel blocking drugs, and isoniazid)	May induce loss of glycemic control in patients.

CONCLUSION

Biguanide derivative metformin HCl has a high solubility, and cannot pass the cell membranes easily by passive diffusion. Therefore, according to BCS, it is classified as a Class 3 compound (high solubility and low permeability). It has a slow and incomplete absorption following oral administ, and hence its bioavailability is low (about 50-60%). Its plasma concentration time profile complies with flip-flop pharmacokinetics due to its slower absorption than elimination. Its binding to plasma proteins is negligible and has a large volume of distribution. Following oral administration, it has a plasma elimination half-life of approximately 6 hours. Metformin is excreted unchanged in urine. It is the most effective, safe and well tolerated antidiabetic used worldwide. It is also recommended for use in cancer and PCOS with no clinical use. Although, it is generally well tolerated, it may cause rare but serious side effects such as lactic acidosis. Further studies are required to develop novel drug delivery systems to overcome its low bioavailability. Increased bioavailability can lead to reduction in administered dose, and decrease the incidence of serious side effects. Also, clinical use of metformin in other disease states such as nephrotoxicity, diabetic nephropathy, cancer and PCOS should be further investigated.

REFERENCES

[1] *Diabetes Atlas 4th Edition*, International Diabetes Federation. 2009. Accessed July 19 https://www.idf.org/e-library/epidemiology-research/diabetes-atlas/21-atlas-4th-edition.html.

[2] *Diabetes Atlas 8th Edition*, International Diabetes Federation. 2017. Accessed July 20 www.idf.org/diabetesatlas.

[3] Chawla, A., Chawla, R., and Jaggi, S. 2016. "Microvasular and macrovascular complications in diabetes mellitus: Distinct or

continuum?" *Indian Journal of Endocrinology and Metabolism* 20(4):546-51.

[4] Bastaki, A. 2005. "Diabetes mellitus and its treatment." *International Journal of Diabetes and Metabolism* 13(3):111-34.

[5] Dostalek, M., Akhlaghi, F., and Puzanovova, M. 2012. "Effect of diabetes mellitus on pharmacokinetic and pharmacodynamic properties of drugs." *Clinical Pharmacokinetinetics* 51(8): 481-99.

[6] Setter, S. M., Iltz, J. L., Thams, J., Campbell, R. K. 2003. "Metformin hydrochloride in the treatment of type 2 diabetes mellitus: a clinical review with a focus on dual therapy." *Clinical Therapy* 25(12):2991-3026.

[7] Wareham, N. J., and O'Rahilly, S. 1998. "The changing classification and diagnosis of diabetes. New classification is based on pathogenesis, not insulin dependence." *British Medical Journal* 317(7155):359-60.

[8] American Diabetes Association 2013. "Diagnosis and classification of diabetes mellitus." *Diabetes Care* 36(Supplement 1):S67-S74.

[9] Wild, S., Roglic, G., Green, A., Sicree, R. et al. 2004. "Global prevalence of diabetes: estimates for the year 2000 and projections for 2030." *Diabetes Care* 27(5):1047-53.

[10] Davidson, M. B., and Peters, A. L. 1997. "An overview of metformin in the treatment of type 2 diabetes mellitus." *The American Journal of Medicine* 102(1):99-110.

[11] *Drugs@FDA: FDA Approved Drug Products*. Accessed July 1 https://www.accessdata.fda.gov/scripts/cder/daf/.

[12] Inzucchi, S. E., Bergenstal, R. M., Buse, J. B., Diamant, M. et al. 2012. "Management of hyperglycemia in type 2 diabetes: a patient-centered approach: position statement of the American Diabetes Association (ADA) and the European Association for the Study of Diabetes (EASD)." *Diabetes Care* 35(6):1364-1379.

[13] Hundal, R. S., and Inzucchi, S. E. 2003. "Metformin: new understandings, new uses." *Drugs* 63(18):1879-94.

[14] Hale, T. W., Kristensen, J. H., Hackett, L. P., Kohan, R. et al. 2002. "Transfer of metformin into human milk." *Diabetologia* 45(11):1509-1514.

[15] *The Merck Index*, 2001. Ed. Budavari, S. 13th edition, p: 998. Whitehouse Station: Merck & Co Inc.

[16] Divakar, P., Kumar, D. P., Praveen, C., Sowmya, C. et al. 2013. "Formulation and in vitro evaluation of liposomes containing metformin hydrochloride." *International Journal of Research in Pharmaceutical and Biomedical Sciences* 4(2):479-85.

[17] European Pharmacopoeia. (2004). *2005 with Supplements 5.1 and 5.2.* 5th ed. Main Volume 5.0. Strasbourg: Council of Europe, p:2003.

[18] Chandira, M., Venkateswarlu, B. S., Shankarrao, J. A., Bhowmik, D. et al. 2010. "Formulation and evaluation of extended release tablets containing metformin HCl." *International Journal of ChemTech Research* 2(2):1320-29.

[19] Wanjari, M. M., There, A. W., Tajne, M. R., Chopde, C. T. et al. 2008. "Rapid and simple RPHPLC method for the estimation of metformin in rat plasma." *Indian Journal of Pharmaceutical Sciences* 70(2):198-202.

[20] Cheng, C. L., and Chou, C. H. 2001. "Determination of metformin in human plasma by high-performance liquid chromatography with spectrophotometric detection." *Journal of Chromatography B Biomedical Science and Applications* 762(1):51-8.

[21] Nirupa, G., and Tripathi, U. M. 2013. "RP-HPLC analytical method development and validation for simultaneous estimation of three drugs: glimepiride, pioglitazone, and metformin and its pharmaceutical dosage forms." *Journal of Chemistry* 2013: 1-8.

[22] Bhamare, P. C. B., Natarajan, S. B., Patil, S., Patil, A. A. et al. 2011. "Development and validation of a precise single stability indicating HPLC method for determinations of Metformin hydrochloride and Fenofibrate, in pure form and in pharmaceutical tablets." *International Journal of PharmTech Research* 3(1):505-15.

[23] Graham, G. G., Punt, J., Arora, M., Day, R. O. et al. 2011. "Clinical pharmacokinetics of metformin." *Clinical Pharmacokinetics* 50(2):81-98.

[24] Crison, J. R., Timmins, P., Keung, A., Upreti, V. V. et al. 2012. "Biowaiver approach for biopharmaceutics classification system class 3 compound metformin hydrochloride using in silico modeling." *Journal of Pharmaceutical Sciences* 101(5): 1773-82.

[25] Center for Drug Evaluation and Research. *Clinical Pharmacology and Biopharmaceutics Review(s).* Accessed June 19 https://www.accessdata.fda.gov/drugsatfda_docs/nda/2012/202270Orig1s000ClinPharmR.pdf.

[26] Chou, C. H. 2000. "Uptake and dispersion of metformin in the isolated perfused rat liver." *Journal of Pharmacy and Pharmacology* 52(8):1011-6.

[27] Rozehnal, V., Nakai, D., Hoepner, U., Fischer, T. et al. 2012. "Human small intestinal and colonic tissue mounted in the Ussing chamber as a tool for characterizing the intestinal absorption of drugs." *European Journal of Pharmaceutical Sciences* 46(5):367-73.

[28] Alvi, M. M., and Chatterjee, P. 2014. "A prospective analysis of co-processed non-ionic surfactants in enhancing permeability of a model hydrophilic drug." *AAPS PharmSciTech* 15(2):339-53.

[29] Elsby, R., Chidlaw, S., Outteridge, S., Pickering, S. et al. 2017. "Mechanistic in vitro studies confirm that inhibition of the renal apical efflux transporter multidrug and toxin extrusion (MATE) 1, and not altered absorption, underlies the increased metformin exposure observed in clinical interactions with cimetidine, trimethoprim or pyrimethamine." *Pharmacology Research and Perspectives* 5(5):1-13.

[30] Kim, D. W., and Park, J. B. 2015. "Development and pharmaceutical approach for sustained-released metformin succinate tablets." *Journal of Drug Delivery Science and Technology* 30:90-9.

[31] Lassoued, M. A., Khemiss, F., Sfar, S. 2011. "Comparative study of two in vitro methods for assessing drug absorption: Sartorius SM

16750 apparatus versus Everted Gut Sac." *Journal of Pharmacy and Pharmaceutical Sciences* 14(1):117-27.

[32] Song, N. N., Li, Q. S., Liu, C. X. 2006. "Intestinal permeability of metformin using single-pass intestinal perfusion in rats." *World Journal of Gastroenterology* 12(25):4064-70.

[33] Tucker, G. T., Casey, C., Phillips, P. J., Connor, H. et al. 1981. "Metformin kinetics in healthy subjects and in patients with diabetes mellitus." *British Journal of Clinical Pharmacology* 12(2):235-246.

[34] Amidon, G. L., Lennernäs, H., Shah, V. P., Crison, J. R. 1995. "A theoretical basis for a biopharmaceutic drug classification: the correlation of in vitro drug product dissolution and in vivo bioavailability." *Pharmaceutical Research* 12(3):413-20.

[35] Cheng, C. L., Yu, L. X., Lee, H. L., Yang, C. Y. et al. 2004. "Biowaiver extension potential to BCS Class III high solubility-low permeability drugs: bridging evidence for metformin immediate-release tablet." *European Journal of Pharmaceutical Sciences* 22(4):297-304.

[36] Murphy, C., Pillay, V., Choonara, Y. E., du Toit, L. C. et al. 2012. "Optimization of a dual mechanism gastrofloatable and gastroadhesive delivery system for narrow absorption window drugs." *AAPS PharmSciTech* 13(1):1-15.

[37] Idkaidek, N. M. 2014. "Interplay of biopharmaceutics, biopharmaceutics drug disposition and salivary excretion classification systems." *Saudi Pharmaceutical Journal* 22(1):79-81.

[38] Bhaskar, R., Bhaskar, R., Sagar, M. K., Saini, V. 2013. "Multivariate chemometric assisted analysis of metformin hydrochloride, gliclazide and pioglitazone hydrochloride in bulk drug and dosage forms." *Advanced Pharmaceutical Bulletin* 3(1):79-84.

[39] Mohamed, O., and Elbashir, A. 2019. "Utilization of 4-Chloro-7-Nittobenzo-2-Oxa-1, 3-Diazol (NBD-CL) as chromogenic reagent for determination of metformin hydrochloride (MET) in pharmaceutical formulation." *Asian Journal of Pharmaceutical Research and Development* 7(1):19-23.

[40] Majidano, S., and Yar Khuhawar, M. 2012. "GC Determination of famotidine, ranitidine, cimetidine, and metformin in pharmaceutical preparations and serum using methylglyoxal as derivatizing reagent." *Chromatographia* 75:1311-17.

[41] Uçaktürk, E. 2013. "The development and validation of a gas chromatography-mass spectrometry method for the determination of metformin in human plasma." *Analytical Methods* 5(18):4723-30.

[42] Goedecke, C., Fettig, I., Piechotta, C., Philippa, R. et al. 2017. "A novel GC-MS method for the determination and quantification of metformin in surface water." *Analytical Methods* 9(10):1580-84.

[43] Shankar, M. B., Modi, V. D., Shah, D. A., Bhatt, K. K. et al. 2005. "Estimation of pioglitazone hydrochloride and metformin hydrochloride in tablets by derivative spectrophotometry and liquid chromatographic methods." *Journal of AOAC International* 88(4):1167-72.

[44] Kolte, B. L, Raut, B. B., Deo, A. A., Bagool, M. A. et al. 2004. "Simultaneous high-performance liquid chromatographic determination of pioglitazone and metformin in pharmaceutical-dosage form." *Journal of Chromatographic Science* 42(1):27-31.

[45] Godasu, S. K., and Sreenivas, S. A. 2017. "A new validated RP-HPLC method for the determination of metformin HCL and empagliflozin in its bulk and pharmaceutical dosage forms." *International Journal of Pharmaceutical Sciences and Research* 8(5):2223-32.

[46] Gulsun, T., and Sahin, S. 2016. "Development and Validation of a RP-HPLC method for simultaneous determination of metformin hydrochloride, phenol red and metoprolol tartrate for intestinal perfusion studies." *Latin American Journal of Pharmacy* 35(1):50-7.

[47] Deng, B., Shi, A., Kang, Y., Li, L. 2011. "Determination of metformin hydrochloride using precolumn derivatization with acetaldehyde and capillary electrophoresis coupled with electro-chemiluminescence." *Luminescence* 26(6):592-7.

[48] Doomkaew, A., Prutthiwanasan, B., Suntornsuk, L. 2014. "Simultaneous analysis of metformin and cyanoguanidine by

capillary zone electrophoresis and its application in a stability study." *Journal of Seperation Sciences* 37(13):1687-93.

[49] Wang, M., and Miksa, I. R. 2007. "Multi-component plasma quantitation of anti-hyperglycemic pharmaceutical compounds using liquid chromatography–tandem mass spectrometry." *Journal of Chromatography B* 856(1):318-27.

[50] Ayoub, B. M., and Mowaka, S. 2017. "LC-MS/MS Determination of empagliflozin and metformin." *Journal of Chromatographic Sciences* 55(7):742-7.

[51] Sharma, K., Pawar, G., Yadam, S., Giri, S. et al. 2013. "LC-MS/MS-ESI method for simultaneous quantitation of metformin and repaglinidie in rat plasma and its application to pharmacokinetic study in rats." *Biomedical Chromatography* 27(3):356-64.

[52] Al Bratty, M., Alhazmi, H. A., Javed, S. A., Lalitha, K. G. 2017. "Development and validation of LC–MS/MS method for simultaneous determination of metformin and four gliptins in human plasma." *Chromatographia* 80(6):891-99.

[53] Oertel, R., Baldauf, J., Rossmann, J. 2018. "Development and validation of a hydrophilic interaction liquid chromatography-tandem mass spectrometry method for the quantification of the antidiabetic drug metformin and six others pharmaceuticals in wastewater." *Journal of Chromatography A* 1556:73-80.

[54] Havel, S. S., and Dhaneshwar, S. R. 2011. "Application of high performance thin layer chromatography-densitometry for the simultaneous determination of metformin hydrochloride and glipizide in bulk drug and tablet formulation." *Der Pharmacia Sinica* 2(1):40-8.

[55] El-Ghobashy, M. R., Yehia, A. M., Mostafa, A. A. 2009. "Application of membrane-selective electrodes for the determination of pioglitazone hydrochloride in the presence of its acid degradant or metformin hydrochloride in tablets and plasma." *Analytical Letters* 42(1):123-140.

[56] Swales, J. G., Gallagher, R. T., Denn, M., Peter, R. M. 2011. "Simultaneous quantitation of metformin and sitagliptin from mouse

and human dried blood spots using laser diode thermal desorption tandem mass spectrometry." *Journal of Pharmaceutical and Biomedical Analysis* 55(3):544-51.

[57] Elgawish, M. S., Nasser, S., Salama, I., Abbas, A. M. et al. 2019. "Liquid chromatography tandem mass spectrometry for the simultaneous determination of metformin and pioglitazone in rat plasma: Application to pharmacokinetic and drug-drug interaction studies." *Journal of Chromatography B* 1124:47-57.

[58] El-Wasseef, D. R. 2012. "Simultaneous determination of metformin, nateglinide and gliclazide in pharmaceutical preparations using micellar liquid chromatography." *International Journal of Biomedical Sciences* 8(2):144-51.

[59] Hassan, S. S. M., Elmosallamy, M. A. F., Mahmoud, W. H., Othman, A. H. M. 1999. "Determination of metformin in pharmaceutical preparations using potentiometry, spectrofluorimetry and UV–visible spectrophotometry." *Analytica Chimica Acta* 378(1):299-311.

[60] El-Bagary, R. I., Elkady, E. F., Ayoub, B. M. 2011. "Spectroflourometric and spectrophotometric methods for the determination of sitagliptin in binary mixture with metformin and ternary mixture with metformin and sitagliptin alkaline degradation product." *International Journal of Biomedical Science* 7(1): 62-9.

[61] Gholivand, M. B., and Mohammadi-Behzad, L. 2013. "Differential pulse voltammetric determination of metformin using copper-loaded activated charcoal modified electrode." *Analytical Biochemistry* 438(1):53-60.

[62] Pentikäinen, P. J., Neuvonen, P. J., Penttilä, A. 1979. "Pharmacokinetics of metformin after intravenous and oral administration to man." *European Journal of Clinical Pharmacology* 16(3):195-202.

[63] Vemula, P., Dodda, D., Balekari, U., Panga, S. 2015. "Simultaneous determination of linagliptin and metformin by reverse phase-high performance liquid chromatography method: An application in quantitative analysis of pharmaceutical dosage forms." *Journal of Advanced Pharmaceutical Technology and Research* 6(1):25-8.

[64] Ahmed, R. 2014. "A simple and convenient method for the simultaneous in vitro study of metformin and glimepiride tablets." *Pakistan Journal of Pharmaceutical Sciences* 27(6):1939-43.

[65] Satheeshkumar, N., Pradeepkumar, M., Shanthikumar, S., Rao, V. J. 2014. "Development of validated stability indicating assay method for simultaneous estimation of metformin hydrochloride and vildagliptin by RP-HPLC." *Drug Research (Stuttg)* 64(3):124-9.

[66] Pandit, V., Pai, R. S., Devi, K., Singh, G. et al. 2012. "Development and validation of the liquid chromatographic method for simultaneous estimation of metformin, pioglitazone, and glimepiride in pharmaceutical dosage forms." *Pharmaceutical Methods* 3(1):9-13.

[67] Sultana, N., Arayne, M. S., Shafi, N., Siddiqui, F. A. et al. 2011. "Development and validation of new assay method for the simultaneous analysis of diltiazem, metformin, pioglitazone and rosiglitazone by RP-HPLC and its applications in pharmaceuticals and human serum." J*ournal of Chromatographic Sciences* 49(10):774-9.

[68] Arayne, M. S., Sultana, N., Zuberi, M. H., Siddiqui, F. A. 2010. "Simultaneous determination of metformin, cimetidine, famotidine, and ranitidine in human serum and dosage formulations using HPLC with UV detection." *Journal of Chromatographic Sciences* 48(9):721-5.

[69] Salem, H. 2010. "Determination of metformin hydrochloride and glyburide in an antihyperglycemic binary mixture using high-performance liquid chromatographic-UV and spectrometric methods." *Journal of AOAC International* 93(1):133-40.

[70] Strugaru, A. M., Kazakova, J., Butnaru, E., Caba, I. C. et al. 2019. "Simultaneous determination of metformin and glimepiride in human serum by ultra high performance liquid chromatography quadrupole time of flight mass spectrometry detection." *Journal of Pharmaceutical and Biomedical Analysis* 165:276-83.

[71] Prasad, P., Satyanaryana, K., Krishnamohan, G. 2015. "Development and validation of a method for simultaneous determination of

metformin and saxagliptin in a formulation by RP-HPLC." *American Journal of Analytical Chemistry* 6:841-50.

[72] Dunn, C. J., and Peters, D. H. 1995. "Metformin. A review of its pharmacological properties and therapeutic use in non-insulin-dependent diabetes mellitus." *Drugs* 49(5):721-49.

[73] Padwal, R. S., Gabr, R. Q., Sharma, A. M., Langkaas, L. A. et al. 2011. "Effect of gastric bypass surgery on the absorption and bioavailability of metformin." *Diabetes Care* 34(6):1295-300.

[74] Sweetman S, ed. (2007). *Martindale: the complete drug reference.* London: Pharmaceutical Press (Electronic version).

[75] Adikwu, M. U., Yoshikawa, Y., Takada, K. 2003. "Bioadhesive delivery of metformin using prosopis gum with antidiabetic potential." *Biological and Pharmaceutical Bulletin* 26(5):662-6.

[76] Boldhane, S. P., and Kuchekar, B. S. 2009. "Gastroretentive drug delivery of metformin hydrochloride: formulation and in vitro evaluation using 3(2) full factorial design." *Current Drug Delivery* 6(5):477-85.

[77] Raparla, R., Talasila, E. G. K. M. 2013. "Design and evaluation of floating drug delivery systems of metformin with natural gums as release retarding polymers." *International Journal of Advances in Pharmaceutics* 1(1):22-8.

[78] Sambol, N. C., Brookes, L. G., Chiang, J., Goodman, A. M. et al. 1996. "Food intake and dosage level, but not tablet vs solution dosage form, affect the absorption of metformin HCl in man." *British Journal of Clinical Pharmacology* 42(4):510-2.

[79] Hilgendorf, C., Ahlin, G., Seithel, A., Artursson, P. et al. 2007. "*E*xpression of thirty-six drug transporter genes in human intestine, liver, kidney, and organotypic cell lines." *Drug Metabolism and Disposition* 35(8):1333-40.

[80] Marathe, P. H., Wen, Y., Norton, J., Greene, D. S. et al. 2000. "Effect of altered gastric emptying and gastrointestinal motility on metformin absorption." *British Journal of Clinical Pharmacology* 50(4):325-32.

[81] Yuen, K. H., Wong, J. W., Billa, N., Julianto, T. et al. 1999. "Bioequivalence of a generic metformin tablet preparation."

International Journal of Clinical Pharmacology and Therapeutics 37(7):319-22.

[82] *Glucophage (Metformin hydrochloride tablets) Label Information.* Accessed August 15 https://www.accessdata.fda.gov/drugsatfda_docs/label/2017/020357s037s039,021202s021s023lbl.pdf.

[83] Klepser, T. B., and Kelly, M. W. 1997. "Metformin hydrochloride: an antihyperglycemic agent." *American Journal of Health System Pharmacy* 54(8):893-903.

[84] Garrison, K. L., Sahin, S., Benet, L. Z. 2015. "Few drugs display flip-flop pharmacokinetics and these are primarily associated with classes 3 and 4 of the BDDCS." *Journal of Pharmaceutical Sciences* 104(9):3229-35.

[85] Robert, F., Fendri, S., Hary, L., Lacroix, C. et al. 2003. "Kinetics of plasma and erythrocyte metformin after acute administration in healthy subjects." *Diabetes and Metabolism* 29(3):279-83.

[86] McCreight, L. J., Stage, T. B., Connelly, P., Lonergan, M. et al. 2018. "Pharmacokinetics of metformin in patients with gastrointestinal intolerance." *Diabetes, Obesity and Metabolism* 20(7):1593-1601.

[87] Najib, N., Idkaidek, N., Beshtawi, M., Bader, M. et al. 2002. "Bioequivalence evaluation of two brands of metformin 500 mg tablets (Dialon & Glucophage)--in healthy human volunteers." *Biopharmaceutics and Drug Disposition* 23(7):301-6.

[88] Vlahov, V., Thyroff-Friesinger, U., Koytchev, R., Bakracheva, N. et al. 2005. "Bioequivalence studies with metformin: comparability of reference tablets from different origins." *International Journal of Clinical Pharmacology and Therapeutics* 43(9):457-62.

[89] Food and Drug Administration. *Waiver of In Vivo Bioavailability and Bioequivalence Studies for Immediate-Release Solid Oral Dosage Forms Based on a Biopharmaceutics Classification System. Guidance for Industry.* 2017. Accessed July 25. https://www.fda.gov/regulatory-information/search-fda-guidance-documents/waiver-vivo-bioavailability-and-bioequivalence-studies-immediate-release-solid-oral-dosage-forms.

[90] European Medicines Agency, *Guideline on the Investigation of Bioequivalence.* 2010. Accessed July 22. https://www.ema.europa.eu/en/documents/scientific-guideline/guideline-investigation-bioequivalence-rev1_en.pdf.

[91] Brookes, L. O., Sambol, N. C., Lin, E. T. et al. 1991. "Effect of dosage form, dose and food on the pharmacokinetics of metformin" *Pharmaceutical Research* 8 Suppl.: S320.

[92] Noel, M. 1980. "Kinetic study of normal and sustained release dosage forms of metformin in normal subjects." *Journal of International Biomedical Information and Data-IBID* 1:9-20.

[93] Rao, N., Chou, T., Ventura, D., Abramowitz, W. 2005. "Investigation of the pharmacokinetic and pharmacodynamic interactions between memantine and glyburide/metformin in healthy young subjects: a single-center, multiple-dose, open-label study." *Clinical Therapy* 27(10):1596-606.

[94] Lee, S. H., and Kwon, K. I. 2004. "Pharmacokinetic-pharmacodynamic modeling for the relationship between glucose-lowering effect and plasma concentration of metformin in volunteers." *Achieves of Pharmacal Research* 27(7):806-10.

[95] Arafat, T., K. A., Shami, M. et al. 1994. "Pharmacokinetics and pharmacodynamics of two oral formulations of metformin hydrochloride." *Advanced Therapy* II:21-33.

[96] Sambol, N. C., Chiang, J., O'Conner, M., Liu, C. Y. et al. 1996. "Pharmacokinetics and pharmacodynamics of metformin in healthy subjects and patients with noninsulin-dependent diabetes mellitus." *Journal of Clinical Pharmacology* 36(11):1012-21.

[97] Sambol, N. C., Chiang, J., Lin, E. T., Goodman, A. M. et al. 1995. "Kidney function and age are both predictors of pharmacokinetics of metformin." *Journal of Clinical Pharmacology* 35(11):1094-102.

[98] Cullen, E., Liao, J., Lukacsko, P., Niecestro, R. et al. 2004. "Pharmacokinetics and dose proportionality of extended-release metformin following administration of 1000, 1500, 2000 and 2500 mg in healthy volunteers." *Biopharmaceutics and Drug Disposition* 25(6):261-3.

[99] Timmins, P., Donahue, S., Meeker, J., Marathe, P. 2005. "Steady-state pharmacokinetics of a novel extended-release metformin formulation." *Clinical Pharmacokinetics* 44(7):721-9.

[100] Food and Drug Administration, *GLUMETZA® - metformin hydrochloride tablet, film coated, extended release*. Accessed July 25 https://www.accessdata.fda.gov/drugsatfda_docs/label/2009/021748s005lbl.pdf.

[101] Ali, S., and Fonseca, V. 2012. "Overview of metformin: special focus on metformin extended release." *Expert Opinion on Pharmacotherapy* 13(12):1797-805.

[102] Briggs, G. G., Ambrose, P. J., Nageotte, M. P., Padilla, G. et al. 2005. "Excretion of metformin into breast milk and the effect on nursing infants." *Obstetrics and Gynecology* 105(6):1437-41.

[103] Vanky, E., Zahlsen, K., Spigset, O., Carlsen, S. M. 2005. "Placental passage of metformin in women with polycystic ovary syndrome." *Fertility and Sterility* 83(5):1575-8.

[104] Scherneck, S., Schlinke, N., Beck, E., Grupe, K. et al. 2018. "Pregnancy outcome after first-trimester exposure to metformin: A prospective cohort study." *Reproductive Toxicology* 81:79-83.

[105] Lipska, K. J., Bailey, C. J., Inzucchi, S. E. 2011. "Use of metformin in the setting of mild-to-moderate renal insufficiency." *Diabetes Care* 34(6): 1431-7.

[106] Scheen, A. J. 1996. "Clinical pharmacokinetics of metformin." *Clinical Pharmacokinetics* 30(5): 359-371.

[107] Food and Drug Administration, *GLUCOPHAGE® (metformin hydrochloride) Tablets, GLUCOPHAGE® XR (metformin hydrochloride) Extended-Release Tablets*. Accessed July 22 https://www.accessdata.fda.gov/drugsatfda_docs/label/2017/020357s037s039,021202s021s023lbl.pdf.

[108] Kusuhara, H., Ito, S., Kumagai, Y., Jiang, M. et al. 2011. "Effects of a MATE protein inhibitor, pyrimethamine, on the renal elimination of metformin at oral microdose and at therapeutic dose in healthy subjects." *Clinical Pharmacology and Therapeutics* 89(6):837-44.

[109] Choi, Y. H., and Lee, M. G. 2006. "Effects of enzyme inducers and inhibitors on the pharmacokinetics of metformin in rats: involvement of CYP2C11, 2D1 and 3A1/2 for the metabolism of metformin." *British Journal of Pharmacology* 149(4):424-30.

[110] Michels, G. M., Boudinot, F. D., Ferguson, D. C., Hoenig, M. 1999. "Pharmacokinetics of the antihyperglycemic agent metformin in cats." *American Journal of Veterinary Research* 60(6):738-42.

[111] Choi, M. K., and Song, I. S. 2008. "Organic cation transporters and their pharmacokinetic and pharmacodynamic consequences." *Drug Metabolism and Pharmacokinetic* 23(4):243-53.

[112] Motohashi, H., and Inui, K. 2013. "Organic cation transporter OCTs (SLC22) and MATEs (SLC47) in the human kidney." *The AAPS Journal* **15**(2):581-8.

[113] Shu, Y., Sheardown, S. A., Brown, C., Owen, R. P. et al. 2007. "Effect of genetic variation in the organic cation transporter 1 (OCT1) on metformin action." *Journal of Clinical Investigation* 117(5):1422-31.

[114] Kashi, Z., Masoumi, P., Mahrooz, A., Hashemi-Soteh, M. B. et al. 2015. "The variant organic cation transporter 2 (OCT2)-T201M contribute to changes in insulin resistance in patients with type 2 diabetes treated with metformin." *Diabetes Research and Clinical Practice* 108(1):78-83.

[115] Müller, J., Lips, K. S., Metzner, L., Neubert, R. H. et al. 2005. "Drug specificity and intestinal membrane localization of human organic cation transporters (OCT)." *Biochemical Pharmacology* 70(12):1851-60.

[116] Song, I. S., Shin, H. J., Shim, E. J., Jung, I. S. et al. 2008. "Genetic variants of the organic cation transporter 2 influence the disposition of metformin." *Clinical Pharmacology and Therapeutics* 84(5):559-62.

[117] Shu, Y., Brown, C., Castro, R. A., Shi, R. J. et al. 2008. "Effect of genetic variation in the organic cation transporter 1, OCT1, on metformin pharmacokinetics." *Clinical Pharmacology and Therapeutics* 83(2):273-80.

[118] Tzvetkov, M. V., Vormfelde, S. V., Balen, D., Meineke, I. et al. 2009. "The effects of genetic polymorphisms in the organic cation transporters OCT1, OCT2, and OCT3 on the renal clearance of metformin." *Clinical Pharmacology and Therapeutics* 86(3):299-306.

[119] Chen, Y., Li, S., Brown, C., Cheatham, S. et al. 2009. "Effect of genetic variation in the organic cation transporter 2 on the renal elimination of metformin." *Pharmacogenet Genomics* 19(7):497-504.

[120] Wang, D. S., Kusuhara, H., Kato, Y., Jonker, J. W. et al. 2003. "Involvement of organic cation transporter 1 in the lactic acidosis caused by metformin." *Molecular Pharmacology* 63(4):844-8.

[121] Higgins, J. W., Bedwell, D. W., Zamek-Gliszczynski, M. J. 2012. "Ablation of both organic cation transporter (Oct)1 and Oct2 alters metformin pharmacokinetics but has no effect on tissue drug exposure and pharmacodynamics." *Drug Metabolism and Disposition* 40(6):1170-1177.

[122] Zhou, M., Xia, L., Wang, J. 2007. "Metformin transport by a newly cloned proton-stimulated organic cation transporter (plasma membrane monoamine transporter) expressed in human intestine." *Drug Metabolism and Disposition* 35(10):1956-62.

[123] Gong, L., Goswami, S., Giacomini, K. M., Altman, R. B. et al. 2012. "Metformin pathways: pharmacokinetics and pharmacodynamics." *Pharmacogenet Genomics* 22(11):820-7.

[124] Ito, S., Kusuhara, H., Yokochi, M., Toyoshima, J. et al. 2012. "Competitive inhibition of the luminal efflux by multidrug and toxin extrusions, but not basolateral uptake by organic cation transporter 2, is the likely mechanism underlying the pharmacokinetic drug-drug interactions caused by cimetidine in the kidney." *Journal of Pharmacology and Experimental Therapeutics* 340(2):393-403.

[125] Lepist, E. I., and Ray, A. S. 2016. "Renal transporter-mediated drug-drug interactions: Are they clinically relevant?" *Journal of Clinical Pharmacology* 56 Suppl 7:S73-81.

[126] Liang, X., Chien, H. C., Yee, S. W., Giacomini, M. M. et al. 2015. "Metformin is a substrate and inhibitor of the human thiamine

transporter, THTR-2 (SLC19A3)." *Molecular Pharmacology* 12(12):4301-10.

[127] Nakamichi, N., Shima, H., Asano, S., Ishimoto, T. et al. 2013. "Involvement of carnitine/organic cation transporter OCTN1/SLC22A4 in gastrointestinal absorption of metformin." *Journal of Pharmaceutical Sciences* 102(9):3407-17.

[128] Ikeda, T., Iwata, K., Murakami, H. 2000. "Inhibitory effect of metformin on intestinal glucose absorption in the perfused rat intestine." *Biochemical Pharmacology* 59(7): 887-90.

[129] McIntyre, H. D., Ma, A., Bird, D. M., Paterson, C. A. et al. 1991. "Metformin increases insulin sensitivity and basal glucose clearance in Type 2 (non-insulin dependent) diabetes mellitus." *Australian and New Zealand Journal of Medicine* 21(5):714-719.

[130] DeFronzo, R. A., Barzilai, N., Simonson, D. C. 1991. "Mechanism of metformin action in obese and lean noninsulin-dependent diabetic subjects." *Journal of Clinical Endocrinology and Metabolism* 73(6):1294-301.

[131] Reaven, G. M., Johnston, P., Hollenbeck, C. B., Skowronski, R. et al. 1992. "Combined metformin-sulfonylurea treatment of patients with noninsulin-dependent diabetes in fair to poor glycemic control." *Journal of Clinical Endocrinology and Metabolism* 74(5):1020-6.

[132] Musi, N., Hirshman, M. F., Nygren, J., Svanfeldt, M. et al. 2002. "Metformin increases AMP-activated protein kinase activity in skeletal muscle of subjects with type 2 diabetes." *Diabetes* 51(7):2074-81.

[133] Rena, G., Hardie, D. G., Pearson, E. R. 2017. "The mechanisms of action of metformin." *Diabetologia* 60(9): 1577-85.

[134] Kanazawa, I. 2011. "A novel attribute of insulin secretion via incretin axis by metformin drug." *International Journal of Endocrinology and Metabolism* 9:420-1.

[135] Maida, A., Lamont, B. J., Cao, X., Drucker, D. J. 2011. "Metformin regulates the incretin receptor axis via a pathway dependent on peroxisome proliferator-activated receptor-alpha in mice." *Diabetologia* 54(2):339-49.

[136] Viollet, B., Guigas, B., Sanz Garcia, N., Leclerc, J. et al. 2012. "Cellular and molecular mechanisms of metformin: an overview." *Clinical Science (Lond)* 122(6):253-70.

[137] Canadian Diabetes Association Clinical Practice Guidelines Expert Committee, Harper, W., Clement, M., Goldenberg, R. et al. 2013. "Pharmacologic management of type 2 diabetes." *Canadian Journal of Diabetes* (37):S61-68.

[138] Jung, S. H., Chae, J. W., Song, B. J., Kwona, K. I., 2014. "Bioequivalence comparison of two formulations of fixed-dose combination glimepiride/metformin (2/500 mg) tablets in healthy volunteers." *Iran Journal of Pharmaceutical Research* 13(2):365-71.

[139] World Health Organization. (2015). *Diabetes*. Accessed August 1. www.who.int/mediacentre/factsheets/fs312/en/.

[140] Beysel, S., Unsal, I. O., Kizilgul, M., Caliskan, M. et al. 2018. "The effects of metformin in type 1 diabetes mellitus." *BMC Endocrine Disorders* 18(1): 1.

[141] Lund, S. S., Tarnow, L., Astrup, A. S., Hovind PLund, S. S. et al. 2008. "Effect of adjunct metformin treatment in patients with type-1 diabetes and persistent inadequate glycaemic control. A randomized study." *PLoS One* 3(10):e3363.

[142] Konrad, K., Datz, N., Engelsberger, I., Grulich-Henn, J. et al. 2015. "Current use of metformin in addition to insulin in pediatric patients with type 1 diabetes mellitus: an analysis based on a large diabetes registry in Germany and Austria." *Pediatric Diabetes* 16(7):529-37.

[143] Johnson, N. P. 2014. "Metformin use in women with polycystic ovary syndrome." *Annals of Translational Medicine* 2(6):56.

[144] Rouabhia, S., Milic, N., Abenavoli, L. 2014. "Metformin in the treatment of non-alcoholic fatty liver disease: safety, efficacy and mechanism." *Expert Review of Gastroenterology and Hepatololgy* 8(4):343-9.

[145] Nasri, H., Rafieian-Kopaei, M. 2014. "Metformin: Current knowledge." J*ournal of Research in Medical Sciences: The Official Journal of Isfahan University of Medical Sciences* 19(7):658-64.

[146] Palomba, S., Falbo, A., Zullo, F., Orio, F. 2009. "Evidence-based and potential benefits of metformin in the polycystic ovary syndrome: A comprehensive review." *Endocrine Reviews* 30(1):1-50.

[147] Rafieian-Kopaei, M., and Nasri, H. 2013. "Ginger and diabetic nephropathy." *Journal of Renal Injury Prevention* 2(1): 9-10.

[148] Hsu, W., Hsiao, P., Lin, P., Chen, S. et al. 2018. "Effect of metformin on kidney function in patients with type 2 diabetes mellitus and moderate chronic kidney disease." *Oncotarget* 9:5416-23.

[149] Cetin, M., and Sahin S. 2016. "Microparticulate and nanoparticulate drug delivery systems for metformin hydrochloride." *Drug Delivery* 23(8): 2796-2805.

[150] Haugrud, A. B., Zhuang, Y., Coppock, J. D., Miskimins, W. K. 2014. "Dichloroacetate enhances apoptotic cell death via oxidative damage and attenuates lactate production in metformin-treated breast cancer cells." *Breast Cancer Research and Treatment* 147(3):539-50.

[151] Anisimov, V. N. 2010. "Metformin for aging and cancer prevention." *Aging (Albany NY)* 2(11):760-74.

[152] Ganjali, M., and Ganjali, H. 2013. "Anticancer effect of metformin, an anti diabetic drug, on breast cancer cells." *Journal of Novel Applied Sciences* 2:796-801.

[153] Wahdan-Alaswad, R., Fan, Z., Edgerton, S. M., Liu, B. et al. 2013. "Glucose promotes breast cancer aggression and reduces metformin efficacy." *Cell Cycle* 12(24):3759-69.

[154] Ioannou, G. N., and Boyko, E. J. 2011. "Metformin and colorectal cancer risk in diabetic patients." *Diabetes Care* 34(10):2336-7.

[155] Nangia-Makker, P., Yu, Y., Vasudevan, A., Farhana, L. et al. 2014. "Metformin: A potential therapeutic agent for recurrent colon cancer." *PLoS One* 9(1):e84369.

[156] Dowling, R. J., Goodwin, P. J., Stambolic, V. 2011. "Understanding the benefit of metformin use in cancer treatment." *BMC Medicine* 9:33.

[157] Kourelis, T. V., and Siegel, R. D. 2012. "Metformin and cancer: new applications for an old drug." *Medical Oncology* 29(2):1314-27.

[158] Evans, J. M., Donnelly, L. A., Emslie-Smith, A. M., Alessi, D. R. et al. 2005. "Metformin and reduced risk of cancer in diabetic patients." *British Medical Journal* 330(7503):1304-5.

[159] Zi, F., Zi, H., Li, Y., He, J. et al. 2018. "Metformin and cancer: An existing drug for cancer prevention and therapy." *Oncology Letters* 15(1):683-90.

[160] Bailey, C. J. 1992. "Biguanides and NIDDM." *Diabetes Care* 15(6):755-72.

[161] DeFronzo, R., Fleming, G. A., Chen, K., Bicsak, T. A. 2016. "Metformin-associated lactic acidosis: Current perspectives on causes and risk." *Metabolism* 65(2):20-9.

[162] Shah, M. A., Niazi, S. G., Tahir, I. M., Shahid, I. et al. 2019. "Correlation of metformin with intravenous iodinated contrast media and precautions." *Pakistan Journal of Pharmaceutical Sciences* 32:859-63.

[163] Chapman, L. E., Darling, A. L., Brown, J. E. 2016. "Association between metformin and vitamin B12 deficiency in patients with type 2 diabetes: A systematic review and meta-analysis." *Diabetes & Metabolism* 42(5): 316-27.

[164] Gilligan, M. A. 2002. "Metformin and vitamin B12 deficiency." *Archieves of International Medicine* 162(4):484-5.

[165] Balsells, M., García-Patterson, A., Solà, I., Roqué, M. et al. 2015. "Glibenclamide, metformin, and insulin for the treatment of gestational diabetes: a systematic review and meta-analysis." *British Medical Journal* 350:h102.

[166] Lalau, J. D., Vermersch, A., Hary, L., Andrejak, M. et al. 1990. "Type 2 diabetes in the elderly: an assessment of metformin (metformin in the elderly)." *International Journal of Clinical Pharmacology, Theraphy, and Toxicology* 28(8):329-32.

[167] lory, J., Haynes, K., Leonard, C. E., Hennessy, S. 2015. "Proton pump inhibitors do not impair the effectiveness of metformin in patients with diabetes." *British Journal of Clinical Pharmacology* 79(2):330-6.

[168] Johansson, S., Read, J., Oliver, S., Steinberg, M. et al. 2014. "Pharmacokinetic evaluations of the co-administrations of vandetanib and metformin, digoxin, midazolam, omeprazole or ranitidine." *Clinical Pharmacokinetics* 53(9):837-47.

[169] Cho, S. K., Kim, C. O., Park, E. S., Chung, J. Y. 2014. "Verapamil decreases the glucose-lowering effect of metformin in healthy volunteers." *British Journal of Clinical Pharmacology* 78(6):1426-32.

[170] Pakkir Maideen, N. M., Jumale, A., Balasubramaniam, R. 2017. "Drug interactions of metformin involving drug transporter proteins." *Advanced Pharmaceutical Bulletin* 7(4):501-5.

[171] Triplitt, C. 2006. "Drug interactions of medications commonly used in diabetes." *Diabetes Spectrum* 19(4):202-11.

[172] Shenoy, C. 2006. "Metformin-associated lactic acidosis precipitated by acute renal failure." *The American Journal of the Medical Sciences* 331(1): 55-7.

In: The Pharmacological Guide to Metformin ISBN: 978-1-53616-634-7
Editor: Natalie Lajoie

Chapter 3

METFORMIN: WHO ARE SUITABLE PATIENTS?

Karel Kubát*

Department of Internal Medicine, Litomerice Hospital,
Litomerice, Czech Republic

ABSTRACT

Metformin is currently the most widely-used drug for the treatment of diabetes mellitus. Perhaps the main therapeutic effect of metformin (not only an unwanted one) is restriction of aerobic metabolism resulting in lower energy income and adenosine triphosphate (ATP) production.

The ultimate effect is increased income of glucose to the cell enabling constant ATP production. This mechanism explains why metformin decreases glucose production by liver cells and restores its uptake by skeletal muscle cells from the blood, even if they do not respond sufficiently to insulin-stimulated impulses. Metformin increases intestinal absorption of glucose, but its turnover into lactate is significantly increased, with intestinal metabolism subsequently becoming ineffective (more anaerobic).

---

* Corresponding Author's E-mail: kkubat.karel@seznam.cz, k.kubat@nemocnice-lt.cz.

The rate of glucose absorption is reduced, and the process occurs more distally along the tract. In the end, metformin decreases net intestinal glucose uptake and aerobic glucose utilization.

However, the side effects of this drug can be dangerous. Metformin-induced lactic acidosis (MILA) is a lethal disorder. The first significant deviation – and probably the cause of MILA development in monitored patients – is gastrointestinal tract involvement. Common infection or the effect of metformin itself cause vomiting and diarrhea, with subsequent hypovolemia causing pre-renal acute kidney injury, which in turn raises metformin levels up to toxic values. Development of kidney failure (followed by heart failure with dyspnea and brain failure with a delirious state) is often unexpected and fast.

Metformin has become the most widely-used drug for the treatment of Type-2 diabetes mellitus. Strictly speaking – we use this drug in cases of treatment of "syndrome of intracellular overfeeding with secondary syndrome of chronic hyperglycemia", today known as "DMT2 syndrome", or "Type-2 diabetes mellitus with obesity". It is the drug of choice for the treatment of T2DM in obese people today. Its use is not recommended in cases of DM with intracellular starvation, i.e., true "Type-1 diabetes mellitus, T1DM (a disease characterized by lack of insulin), but can be useful in patients with either type of diabetes mellitus, i.e., both T1DM and T2DM.

Many patients with intracellular starvation due to a lack of insulin (T1DM) tend to overfeed – then treated with high, supra-normal, doses of insulin subsequently develop intracellular overfeeding and, hence, T2DM. On the other hand, many obese patients with T2DM lose their ability of insulin overproduction and they develop ***either*** type of diabetes mellitus.

Today, metformin belongs to the safest medications in its class. Nevertheless – if we are forced to use drugs that limit the surplus of energy and decrease the effectiveness of metabolism – we can still expect an adverse effect characterized by aerobic metabolism collapse, i.e., MILA. Today we have some other modern drugs: incretins acting by the anorectic effect; gliflozins modulating direct glucose elimination via the urine. While adequate food intake and physical training have the same beneficial effect, they are definitely not associated with any potential harm.

Keywords: lactic acidosis, metformin, Type-2 diabetes mellitus, T2DM, AMP-activated protein kinase, respiratory complex I

1. Introduction

Metformin is currently a widely-used drug in the treatment of Type-2 diabetes mellitus (T2DM).

The drug reduces the efficiency of metabolism of energy-rich substances. This is the basic effect of the drug.

Metformin allows patients to receive more energy-rich substances than their cells actually need. It increases the need for glucose by tissue (cells), reduces glucose production thereby lowering blood glucose levels.

The effect of metformin is similar to that of other biguanides (phenphormin and buformin) and is quite different from – and in many aspects opposed to – the effect of insulin.

Unlike insulin, metformin provides protection to the body from the accumulation of energy-rich substances. Besides, it protects the body from obesity as well as steatosis of the liver and muscles including the myocardium.

The main therapeutic effect of the drug is reduction of aerobic metabolism.

In our center specializing, among other things, in the management of acid-base disturbances, we have encountered, over the last 30 years, a relatively large number of cases of biguanide intoxication. These included 31 cases of lactic acidosis caused by the use of buformin in the 1989–1998 period [1] with another 23 cases of metformin-induced lactic acidosis (MILA) between 1999 and 2017 [2], followed by 4 cases (2017–2019) (unpublished data).

The symptoms of intoxication with metformin and buformin were similar. In contrast, they were quite different from those of an insulin overdose. It was for this reason that we decided to revise some of the traditional views on the effect of metformin.

2. VARIOUS EXPERIMENTAL RESULTS

Experiments with metformin bring very ambiguous results. One may ask why?

Metformin exerts its effect at sites where the energy is acquired (transformed). It acts in the mitochondria, where it modulates production of the basic cell energy carrier – adenosine triphosphate (ATP). To reach the mitochondria, it is necessary to cross three membranes, the cell membrane and the outer and inner membranes of the mitochondria.

For metformin to exert its effect, a high concentration at the site of action is required; this is accomplished by either a high dose of metformin or its long-term administration. This is the only way to allow sufficient transfer of the drug across the membranes and to achieve the necessary high concentration in the mitochondrial matrix.

The other two biguanides, buformin and phenformin, are (unlike metformin) effective at significantly lower concentrations. Phenformin is lipophilic, readily crossing cell membranes and penetrating significantly faster to its site of action. This, on the one hand, makes it easy to determine the effect of this substance but, on the other, it is responsible for its high toxicity index.

In the body, the metformin molecule is in ionized form, carrying an electric charge. It is hydrophilic and does not pass rapidly through cell membranes.

For this reason, unfortunately, studies using inadequately low metformin concentrations as well as those using inadequately short observation period reported confusing and occasionally contradictory results.

How high is then the necessary effective concentration of the drug and how long does it take to evaluate the effect of metformin?

Compare:

The plasma concentrations of metformin found in patients are about 20.9 μM (= μmol/l), e.g., 2.7 mg/l [3].

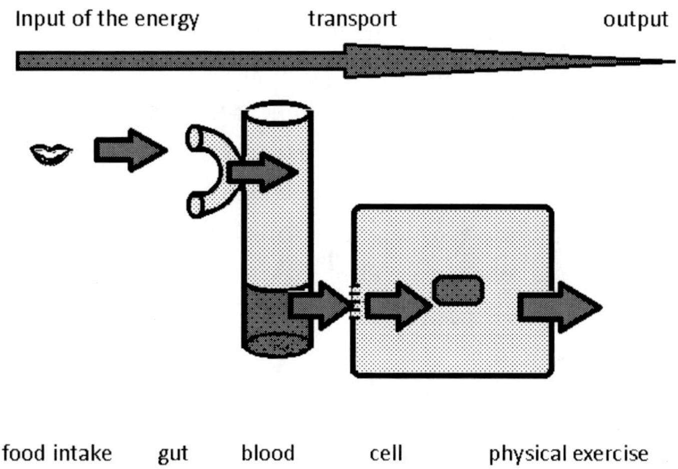

Figure 1. Transport of the energy. Eating is the first principal step of energy transport. Food enters the mouth, passes through intestine to blood. Intracellular metabolism, determine the need of glucose. Physical exercise is the last step of energy transport in the muscle [135].

In experiments the concentrations needed for metabolic changes to occur, in tissue homogenates containing disrupted mitochondria, were 30 000 μM [4]. The concentrations required for 50% inhibition of nicotinamide adenine dinucleotide, NADH oxidation by metformin is 79 000 μM in the case of submitochondrial particles [5].

Thus, the necessary concentration at the site of action is a thousand times higher than the plasma concentrations.

3. Intracellular Metabolism

The most important metabolic processes take place inside the cells. These processes are controlled by external (extracellular) hormones and, (mainly), by internal regulatory mechanisms, through "intracellular hormones". The basic metabolic processes, in which energy is obtained, take place even in the inner organelles inside the cells, in the mitochondria. Similarly· the metabolic processes occurring in the mitochondria are controlled primarily by intramitochondrial regulatory mechanisms through what is referred to as “intramitochondrial hormones”.

While the hormones determine the amount of energy-rich substances (glucose, triacylglycerols, or amino acids) entering the cells· intracellular regulatory mechanisms determine how many energy-rich substances enter the mitochondria. Metabolism is linked to the presence of substrates that actually reach the mitochondria and of the effect of respiratory complexes that produce the bulk of the major energy carrier, adenosine triphosphate (ATP). It is necessary to produce a certain constant amount of ATP. If we reduce the efficiency of the (aerobic) metabolism, then more glucose – that the cell will consume – will be needed. Thus, it takes more glucose from the blood allowing to reduce hyperglycemia.

Why is it not possible to meet the increased need for ATP and, consequently, the need for glucose by increasing physical activity, muscle exercise? The answer is that people today are generally reluctant to exercise physically. On the very contrary, they are keen on eating more food than their body actually needs. And the imbalance between low energy expenditure and high energy intake is solved through medication.

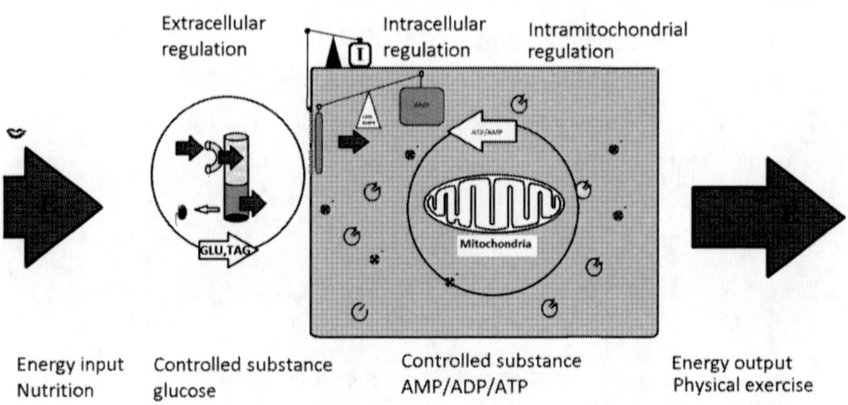

Figure 2. Regulation. 1/ Extracellular regulatory mechanisms - hormonal system. (insulin, glucagon, growth hormone, adrenaline, leptin, adiponectin and others). 2/ Intracellular regulation mechanism (AMPK, LKB1, CaMKK and others). 3/ Intra-mitochondrial regulatory mechanism.

4. Inhibition of Respiratory Complex I in Submitochondrial Particles

As mentioned above, the basic effect of metformin consists in the inhibition of mitochondrial respiratory complex I [4-7]. This complex is located on the inner side of the mitochondria. To assess the effect of metformin, it is necessary to obtain sub-mitochondrial particles (SMPs). To do so, tissues were homogenized and sonicated to disrupt all cells and mitochondria.

In this process, the pH gradient disappears (the pH within the mitochondria is 7.6, while the intracellular pH is 7.0), a fact limiting our ability to accurately determine the effect of other substances or pH to some extent.

Electron transfer by respiratory complex I was partially blocked even at extremely high concentrations of metformin. Metformin inhibited complex I in sonicated tissue homogenates contained disrupted mitochondria at concentrations of 30 000 μmol/l. [3] whereas the concentrations needed for 50% inhibition of NADH oxidation by metformin was 79 000 μmol/l in case of submitochondrial particles [5].

Restricted energy income, restricted ATP production and increasing AMP concentrations all influence the basic mechanism which controls the acquisition and conversion of energy by the cell.

(By the way, the concentrations required to achieve 50% inhibition of NADH oxidation by phenformin is significantly lower, 2230 umol/l, in the case of submitochondrial particles [11].)

5. Inhibition of Respiratory Complex I in Intact Mitochondria

To investigate the effect of metformin on intact mitochondria, not extremely high concentrations but extremely long periods of time are necessary. Metformin is slowly transmitted through membranes by specific carriers. This process is very time consuming. While metformin at concentrations of 10 000 μmol over a short period of time (30 minutes) did not inhibit oxygen consumption of isolated liver mitochondria [8], this agent at relatively low concentrations (50 μM) caused significant inhibition after a long time period (24 and 60 hours) [5] (in permeabilized hepatoma cells).

To be more specific, metformin (50 μM) caused 12.6% inhibition after 24 hours, whereas, for 100 μM, the inhibition rate was 25.8%. At 60 hours, metformin (50 μM) caused 29.4% inhibition, whereas for 100 μM metformin, the inhibition rate was 37.1% [4]. Similar results were obtained in experiments using the rat soleus muscle [4]. Experiments designed to assess the effect of short periods (e.g., 20 minutes) of metformin incubation permeabilized cells showed no effect of metformin on respiratory function [9].

The time and concentration dependence of complex-1 inhibition by metformin in the mitochondria and SMPs suggests a slow rate of permeation of the drug across the inner mitochondrial membrane to directly inhibit complex I [5]. The positive charge on metformin may account for its accumulation within the matrix of energized mitochondria driven by the membrane potential (Δψ). Thermodynamic considerations predict that, with a Δψ of -180 mV, metformin should accumulate 1000-fold within the mitochondrial matrix [5]. Metformin probably helps to avoid life-threatening complications because of its (partially) self-limiting inhibition. As the matrix concentrations of the drug rise, progressive inhibition of the respiratory chain will lead to a drop in the membrane potential, thereby preventing further accumulation of the drug [5].

6. Inhibition of Respiratory Complex I in Intact Cells

Preparation of the cells (we do to be able to evaluate intracellular or intramitochondrial processes) is associated with significant cell damage.

Digitonin permeabilization, which leads to cell membrane disruption, permits the penetration of ions from cells and vice versa. However· the huge concentration differences between sodium, potassium, calcium and pH levels are essential to maintain the vital processes within the cells. Also, breakage of the mitochondrial membrane e.g., by sonification, results in a loss of concentration gradients, pH among other things, which are responsible for gaining energy by these organelles. Reachable data come from experiments obtained under completely non-physiological conditions incompatible with life. Therefore, any processes that require a longer duration of experiment are very difficult to evaluate. The results we get with living intact cells are therefore thus even more important.

In experiments with live undamaged cells, the researcher is faced with another problem. Cells respond to any significant deviation in vital functions, hence processes whereby energy is acquired are controlled by a number of parallel control and performance mechanisms.

Imagine this situation: Metformin will reduce ATP production (and increase the concentrations of energy-poor AMP). The regulatory mechanism detects this dangerous deviation and immediately changes energy metabolism. Regulatory mechanisms limit the formation of energy-rich substances such as cholesterol. Furthermore, regulatory mechanisms increase the ATP production from fatty acids, glucose or amino acids. Regulatory processes very effectively restore ATP (and thus also ADP and AMP) levels to previous values. As a result, an observer, while noticing changes in metabolism but will remain unaware of any significant changes in the AMP/ATP ratio.

First, let us mention the results of experiments where metformin was used at a relatively high concentration. The effect on energy metabolism of intact cell is relatively more intensive and faster than that in permeabilized cells. In intact hepatocytes, metformin 10 000 μM (i.e., 10 mM) decreases the mitochondrial membrane potential ($\Delta\Psi$) by approximately 30 mV after 30 minutes of incubation (DC -175… -141 mV). It likewise reduces the cytosolic ATP/ADP ratio (from 7.3 to 2.5) and mitochondrial ATP/ADP ratio (from 1.7 to 0.5) while increasing the lactate/pyruvate ratio by 23% (from 19.8 to 24.4) and 3-hydroxybutyrate/acetoacetate ratio by 111%

(from 0.18 to 0.38) and decreasing the cellular respiratory rate (JO2) (from 100% to 54%) [8].

Incubation with a lower concentration (2000 μM) of metformin for a longer period of time (3 hours) increases the ratio of β-hydroxybutyrate/acetoacetate by 249% and that of L-lactate/pyruvate by 166% [5]. Similar results were obtained with the isolated rat soleus muscle: its 24-hour exposure to metformin (at a lower concentration) reduced cell respiratory and increased anaerobic glycolysis rates, with 270 μmol metformin increasing lactate production by 84% and decreasing oxidation of glucose [4].

Is it possible to register deviations caused by low metformin concentrations? Assessment of the effect of metformin on the amounts of AMP or ATP in intact cells is relatively difficult. The majority of ATP/ ADP/AMP is processed intramitochondrially. Regulatory mechanisms are set to respond to minor deviations rather than those that cause noticeable changes in the total ATP/AMP concentration in the cell.

The total AMP content in a well perfused rat heart is approx. 350 μM, with cytosolic AMP content being about 0.4 μM [6]. Cytosolic [ADP] and [AMP] represent the metabolically active fractions of the total content of these two nucleotides. In metformin-perfused hearts, cytosolic [AMP] in 10 mM metformin-perfused hearts increased significantly, this, however, after a long period of time (42–50 minutes) [10].

7. Intracellular Regulation

Each cell has its own regulatory system which maintains its energy homeostasis. We are looking for a single molecule that controls the cell's energy status. One molecule cannot control such a complex regulatory mechanism. There are a number of molecules that monitor and affect a cell's energy status. The number is likely to be as extensive as the number of regulatory molecules in the extracellular space. The same as tens or hundreds of extracellular hormones are currently known, tens or hundreds

of intracellular signaling and executive molecules are most likely to be gradually identified.

Which molecules are considered the most important today?

An important energy sensor and regulator of energy homeostasis is AMP-activated protein kinase, AMPK. In the presence of AMP, the activity of AMPK is increased approximately 4–5-fold. AMP or ADP can directly bind to the γ regulatory subunits of AMPK, leading to a conformational change that protects activating phosphorylation of AMPK [11, 12]. AMPK phosphorylation results in its activation by at least 100-fold. AMPK is phosphorylated by upstream AMPK kinases, i.e., AMP-activated protein kinase kinase, AMPKK [13]. The following two molecules are considered to be significant: LKB1 and CaMKKB. However, LKB1 has several superior kinases: protein kinase C ZETA, PKC-ζ [14], which is important for the metformin effect, and another two protein kinases A, PKA and p90 ribosomal S6 kinase, RSK [14]. Another potential mechanism for AMPK activation is through TGFβ-activated protein kinase (TAK-1). [15]. Thus, the central energy sensor is a whole network of upstream kinases. Are these superior kinases controlled by AMP or ADP or ATP? The most likely answer is yes.

AMPK can be phosphorylated in response to calcium flux via the calmodulin-dependent protein kinase kinase-beta, CaMKKβ [16-19].

Liver kinase B1, LKB1, otherwise called serine-threonine kinase 11 (STK11) is an important upstream kinase of adenosine monophosphate-activated protein kinase [20]. Active LKB1 kinase is actually a complex of three proteins: LKB1, Ste20-related kinase adaptor alpha (STRADA) and calcium-binding protein 39 (CAB39; also known as mouse protein 25, MO25). Thus, the enzyme complex is sometimes referred to as LKB1-STRAD-MO25.

LKB1 can phosphorylate AMPK (and some other members of this AMPK subfamily) increasing their activity >50-fold [21].

Later, other AMPK-like enzymes have been identified: BRSK1, BRSK2, NUAK1, NUAK2, QIK, QSK, SIK, MARK1, MARK2, MARK3, MARK4, and MELK. They are called AMPK-related kinase RK family enzymes [20, 21]:

- BRSK1 and BRSK2 are brain-specific serine/threonine kinases 1 and 2, respectively.
- NUAK1 refers to nuclear AMPK-related kinase – AMPK-related kinase 5 (ARK5).
- NUAK2 is also called SNF1/AMPK-related kinase (SNARK).
- SNF1 is sucrose non-fermenting.
- SIK stands for salt-induced kinase. SIK2 was identified as a SIK1-related gene expressed in adipose tissue. SIK3 is also known as QSK.
- MARK1–4 are microtubule affinity-regulating kinases 1 through 4.
- MELK is maternal embryonic leucine zipper kinase.

Which of these molecules are significant or even more significant than AMPK? When will we define all their possible interactions? And are they important for the development of diabetes? Certainly they are important. Long-term research is warranted to assess all their possible effects.

8. AMPK Pathway

5´AMP- activated protein kinase or AMPK is a multi-subunit enzyme which plays the essential role in cellular energy homeostasis. AMPK induces catabolic processes and enables the production of ATP.

It regulates the cellular uptake of glucose, beta-oxidation of fatty acids and biogenesis of the GLUT4 glucose transporter. AMPK increases the glucose input in cases of lack of energy, intracellular starvation, as well as high concentrations of AMP and low ATP concentrations [22-24]. It is a regulator of the lipid biosynthetic pathways due to its role in the inactivation of key enzymes such as acetyl-CoA carboxylase (ACC) [25]. AMPK inhibits anabolic processes that consume ATP, such as protein synthesis [26]. AMPK is a major signaling protein that represses the rate of protein synthesis [27].

AMPK acts as an ultrasensitive cellular energy sensor system responding to increases in AMP concentrations [AMP] or the AMP/ATP ratio [25].

The original idea was simple: AMPK recognizes a lack of energy-rich ATP and an excess of energy-poor AMP molecule to change cell metabolism and restore the equilibrium with sufficient ATP. Later studies showed that the response to the AMP/ATP ratio directed by AMPK alone is relatively small. Superior kinases (LKB1, CaMKKβ and others) have a significant effect. For example: a deficiency of LKB1 in the skeletal muscle attenuates AMPK activation and glucose uptake during muscle contraction [28].

All substances that influence the effect of superior kinases are not yet known. Extracellular signals transmitted from extracellular fluid through hormones and their receptors (e.g., through insulin, leptin, and adiponectin) reflect the needs of the whole body whereas intracellular signal (AMP, ATP or calcium and other substances) express the requirements of a single cell.

An explanation for the effect of calcium and CaMKKβ seems to be relatively easy. The superior kinase CaMKKβ responds to elevated calcium concentrations as a manifestation of pump failure that removes calcium from the cell. The intracellular calcium concentrations are 100 times lower than those in extracellular fluid. High calcium concentrations may indicate targeted opening of calcium channels, necessary for actin and myosin interaction allowing for muscle contraction. Energy in the form of ATP will be needed to relax the muscle. Therefore, calcium concentrations in a cell's cytoplasm is a significant signal of the need for energy supply that is transmitted via calmodulin-dependent protein kinase kinase-beta, CaMKKβ, to AMPK.

The cells respond to very small variations in the plasma AMP/ATP ratio, which are not yet laboratory detectable. Conversely, it is these regulatory mechanisms that we observe and misinterpret as a putative effect of the drug.

Under conditions of severe stress, when the levels of AMP might approach those of ADP and free ATP, the 10-fold allosteric activation caused by AMP would multiply with the >200-fold activation caused by Thr 172 phosphorylation to yield a >1000-fold activation overall. Thus, the mammalian kinase can respond over a wide range of the cellular ADP/ATP and AMP/ATP ratios and over a very wide dynamic range [29].

AMPK has three subunits. Different isoforms exist for each of the AMPK subunits. Two α (α1 and α2), two β (β1 and β2), and three γ (γ1, γ2, and γ3) isoforms enable the creation of up to 12 distinct AMPK configurations. However, in human skeletal muscle, these configurations are likely limited to α2/β2/γ1 (most abundant), α2/β2/γ3, and α1/β2/γ1 [30]. Of these three, α2/β2/γ3 accounts for the majority of AMPK activation due to high-intensity exercise [31].

Metformin activates AMP-activated protein kinase. Directly or indirectly? Via AMP concentrations or via AMPK phosphorylation? Either way?

Some drugs – such as metformin or pioglitazone – affect ATP production. In a system that would not have regulatory mechanisms, a variation in ATP concentration would arise.

Regulatory events in the living cell (maintaining cell homeostasis) alleviate immediately deviation of the ATP concentration and thus deviation of the AMP/ATP ratio. The deviations that occur in a living organism are so small that they are very difficult to detect. Besides, the total amount of ATP and AMP is largely stored in the mitochondria, with only a small portion of these compounds, which are present extra-mitochondrially in the cytoplasmic space, affecting cell metabolism. For this reason, intracellular cytosolic concentrations (not total concentrations) of ATP and ADP should be evaluated. Nevertheless, it was possible to prove the effect of the drug on the energy of living cells: AMPK activity responds to cytosolic concentration ([AMP] hereinafter), e.g., in the isolated rat heart [32, 33].

In these experiments, metformin activated AMPK in the heart by increasing *cytosolic* AMP concentrations [10]. In hearts treated with 10 mM metformin, [AMP] was increased (after a relatively short period, 50 min) and AMPK activity, phosphorylated AMPK, and phosphorylated acetyl-CoA carboxylase were elevated [10]. Whereby *total* ATP content and *total* AMP/ATP ratio remained unchanged or the difference was too small to detect.

Metformin is able to increase the net phosphorylation of the AMPK catalytic α subunit at Thr-172, a crucial phosphorylation site in the activation of AMPK, and activate AMPK activity in primary hepatocytes [34]. However upstream kinase LKB1 is necessary for this process to set in: hepatic knock-out of LKB1 (liver kinase B1), the upstream kinase for AMPKα phosphorylation at Thr-172, abolishes metformin suppression of hepatic glucose production [35]. Furthermore, activation of AMPK results in the phosphorylation of CRTC2 (CREB-regulated transcription coactivator 2) and CBP (CREB-binding protein), which then inhibit gluconeogenic gene expression [36, 37].

Metformin increases the translocation of LKB1S from the nucleus to cytosol. When inactive, LKB1 is localized to the nucleus and is able to translocate to the cytosol and become active when it associates with STRAD and MO25 [38].

Nucleocytoplasmic translocation of LKB1L, which requires phosphorylation of Ser-428/431 and Ser-307 in the LKB1L C-terminal domain, is needed for metformin-mediated activation of AMPK [39, 40].

9. Transport

Metformin is transported by organic cation transporters (OCTs) (solute carrier 22, SLC22) OCT1, OCT2 [41-43] and other polyspecific organic cation transporters.

The following transporters are involved in the drug's absorption, OCT3 (SLC22A3), PMAT (SLC29A4) [44], SERT (SLC6A4), distribution, OCT1 (SLC22A1) OCT3, and elimination OCT2 (SLC22A2), MATE1 (SLC47A1), and MATE2 (SLC47A2) [45].

OCT1 has been shown to play an important role in hepatic metformin uptake, which is an important mechanism for its therapeutic effect as well as side effects (lactic acidosis) [42].

OCT2, which is expressed on the basolateral membrane of kidney tubular cells, was implicated in the renal excretion of metformin [43, 46].

OCT3 is distributed in brain tissue, in the liver, kidney, intestine, and other organs.

PMAT, proton-activated organic cation transporter, or plasma membrane monoamine transporter, is expressed in the human small intestine and concentrated on the tips of the mucosal epithelial layer. PMAT transports many organic cations and also plays a role in the intestinal uptake of metformin [44].

Metformin is also a substrate for and an inhibitor of the human thiamine transporter, THTR-2 (SLC19A3), which is highly expressed in the small intestine [47].

Genetic disorders, drug interactions or pH changes influence the function of transporters. The transport activity of the OCTs is generally significantly decreased at lower pH [48-50], and the acidic environment in the gut lumen (pH 4.0–7.0) may limit the ability of OCT3 in drug absorption. PMAT-mediated metformin transport is greatly stimulated by an acidic pH, with the uptake rate being ~4-fold higher at pH 6.6 than at pH 7.4 [51].

10. THIAMINE

Selective thiamine deficiency is just another possible mechanism for metformin therapeutic (as well as adverse) effects.

Vitamin B1 or thiamine, is essential for normal glucose metabolism. Thiamine pyrophosphate is a cofactor for pyruvate dehydrogenase (PDH), α-ketoglutarate dehydrogenase, and transketolase, all of which have fundamental roles in regulating cellular metabolism [52].

Pyruvate dehydrogenase is a multienzyme complex in the inner mitochondrial membrane that catalyzes the oxidative decarboxylation of pyruvate to acetyl coenzyme A (CoA), which can then enter the citric acid cycle. In thiamine deficiency, when pyruvate cannot undergo this conversion, it is converted to lactate by the action of lactate dehydrogenase [53].

OCT1, the major hepatic transporter for metformin, is also the primary hepatic uptake transporter for thiamine [54].

The human thiamine transporter 2 THTR-2 mediates thiamine uptake in the intestine, CNS, and various peripheral tissues. Metformin inhibits the THTR-2-mediated uptake of thiamine [47].

The gastrointestinal syndrome of thiamine deficiency consisting of nausea, vomiting, abdominal pain and lactic acidosis – gastrointestinal beri-beri – resembles metformin intoxication [55-57].

11. Mitochondrial Glycero-3-Phosphate Dehydrogenase Inhibition

The glycero-3-phosphate shuttle has an important role in the regulation of hepatic gluconeogenesis [58, 59].

It is a mechanism that regenerates NAD^+ from NADH. Cytoplasmic glycero-3-phosphate dehydrogenase 1 (GPDH-C) converts dihydroxyacetone phosphate to glycerol-3-phosphate, by oxidizing one molecule of NADH to NAD^+. Glycero-3-phosphate gets converted back to dihydroxyacetone phosphate by inner membrane-bound mitochondrial glycero-3-phosphate dehydrogenase 2 (GPDH-M or mGDP), reducing one molecule of enzyme-bound flavin adenine dinucleotide (FAD) to $FADH_2$ [60]. $FADH_2$ subsequently reduces coenzyme Q (ubiquinone to ubiquinol) which enters the oxidative phosphorylation pathway. [60]. The net gain is 2 ATP molecules [60].

Metformin non-competitively inhibits the redox shuttle enzyme mGPD, resulting in an altered hepatocellular redox state, reduced conversion of lactate and glycerol to glucose, and decreased hepatic gluconeogenesis [61].

12. Metformin Reduces Liver Glucose Production by Inhibition of Fructose-1-6-Bisphosphatase

Hepatocytes are equipped with a mechanism to control the rate of hepatic gluconeogenesis in response to energy status, and fructose bisphosphatase 1 (FBP1) has long been recognized as a key component [62]. FBP1 catalyzes the irreversible hydrolysis of fructose-1,6-bisphosphate (F-1,6-P2) to fructose-6-phosphate (F6P) and inorganic phosphate (Pi) in the presence of divalent cations. FBP1 activity is regulated synergistically by the allosteric inhibitors AMP and F-2,6-P2. Metformin induces a mild energy stress in the liver, leading to an increase in the concentrations of AMP that allosterically inhibits FBP1 to lower hepatic glucose production (HGP). The AMP-inhibited fructose-1,6-bisphosphatase-1 is also considered to be the major contributor to metformin's therapeutic action (glucose-lowering effect) [63].

12.1. Other Theories

Recently, another study reported that metformin inhibits HGP through hepatic AMPK-independent mechanisms, by attenuating the ability of glucagon to increase 3',5'-cyclic adenosine monophosphate (cAMP) levels and promote HGP [64].

13. Aerobic Metabolism Inhibition Caused by Metformin

What do all the theories that explain the effect of metformin in the body have in common? All theories show that metformin reduces the efficiency of metabolism and increases glucose consumption for basic metabolic processes. Metformin reduces the processes that lead to aerobic metabolism of substances that are rich in energy and contain carbon (especially carbohydrates, fats and amino acids).

The end product of glucose metabolism is not CO_2 +H_2O (with 38 (32 more exactly) ATP molecules being obtained) but lactate (with a 2 ATP molecule gain). The need for glucose in the intestine or muscle increases.

Lactate is produced under physiological conditions, too. At rest, it is formed mainly by the skin, muscle, brain, red blood cells or the intestine. During heavy exercise, the most important source of lactate is white skeletal muscle. Placenta during pregnancy or lung in sepsis produce excessive amounts of lactate. Approximately two thirds of lactate are metabolized in the liver and one third in the kidney.

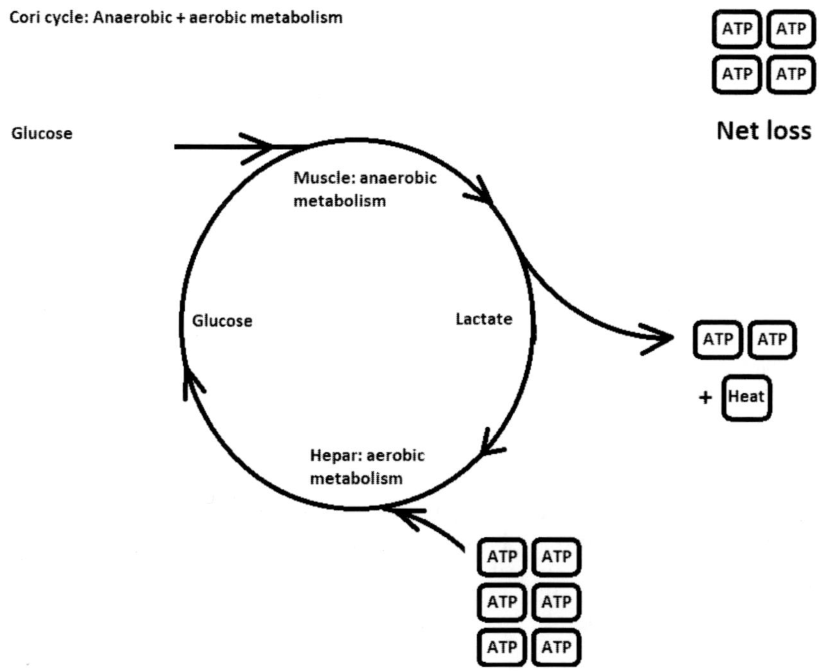

Figure 3. Cori cycle. The Cori cycle (Lactic acid cycle). Lactate produced by anaerobic glycolysis in the muscles moves to the liver and is converted to glucose. Glucose returns to the muscles and is cyclically metabolized to lactate and back.

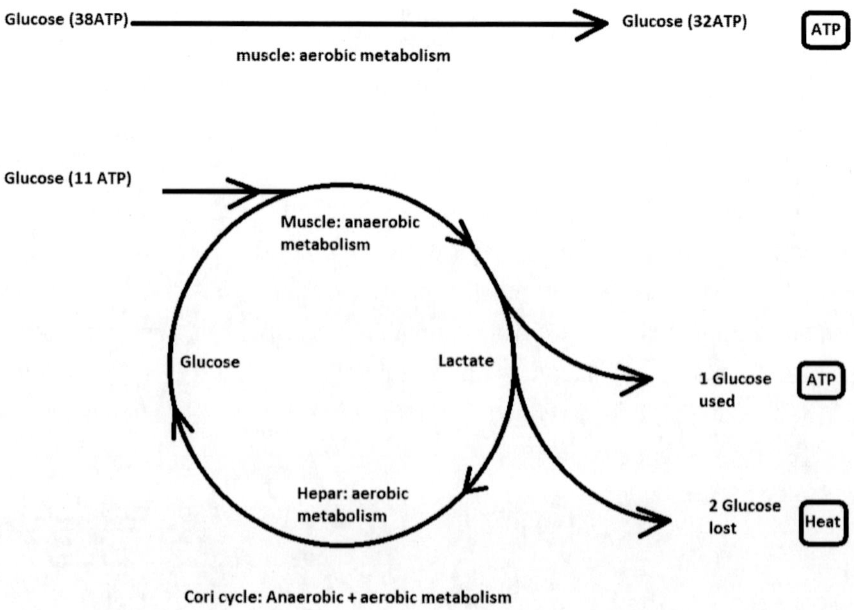

Figure 4. Effectiveness of the Cori cycle. Theoretical metabolic efficiency of the aerobic metabolism: 38 ATP molecules per 1 glucose. Real efficiency of aerobic metabolism: about 32 ATP per 1 glucose, 84% effectiveness (due to leaky membranes or cost of transport into the mitochondrial matrix). Efficiency of anaerobic metabolism (lactate lost by kidney): 2 ATP per 1 glucose, 5.3%. Efficiency of combined metabolism, anaerobic + aerobic in the Cori cycle (lactate converted back to glucose in the liver) 11 ATP per 1 glucose, 33%. 4 molecules ATP are lost every cycle.

It can get converted into glucose (or into CO_2 and H_2O if ATP is needed and aerobic metabolism possible). Metformin has been shown to drive lactate production. It is worth noting that the lactate-glucose-lactate cycle (or glucose-lactate-glucose cycle) is a highly energy-consuming process. Lactate enters the Cori cycle: it is transmitted from the site of origin (muscle, intestine) to where glucose is produced from it (liver), but at the cost of losing 4 ATP molecules (!) to recover each glucose molecule. Metformin causes a futile intestinal-hepatic cycle [65] (and other futile, muscle-hepatic and muscle-kidney cycles) thus increasing energy expenditure [65].

Metformin reduces ATP production with a larger amount of energy-rich compounds consumed by cells to meet the body's metabolic requirements. Glucose, triglycerides and fatty acids are consumed and do not accumulate in the blood.

14. The Intestine

1) Metformin increases intestinal glucose absorption. 2) Metformin lowers intestinal glucose gain. Which of these statements is correct? Both.

Metformin would stimulate glucose uptake from the intestinal lumen to its tissue, primarily in the proximal jejunum – if it were possible – but, in fact reduces transport of glucose to the blood. Enterocytes, which are responsible for glucose entry into the body, highly accumulate metformin. Metformin significantly reduces the metabolism of these cells. Enterocytes themselves consume increased amounts of glucose. This mechanism explains why less glucose is transferred from the intestine to the liver in metformin-treated patients. Blood glucose levels are significantly reduced depending on the dose of metformin.

14.1. Re 1) Metformin Increases the Rate of Glucose Absorption in the Human Intestine

Metformin really increases glucose uptake and utilization [66, 67]. Metformin has been shown to cause an increase in labeled (intravenously administered) glucose uptake (18-fluoro-deoxy-glucose, 18F-FDG) in the colon and small intestine on the PET-CT scan [68-71]. It is critical to withdraw metformin for 2 days before an 18F-FDG PET/CT scan is performed for intra-abdominal neoplasic lesion assessment [70].

Physiological focal uptake – sometimes intense – has been reported within the right lower quadrant, corresponding to the cecum and right colon; it was suggested to be related to the high concentrations of glucose-metabolizing lymphatic cells in this particular region [72].

14.2. Re 2) Metformin Reduces the Amount of Glucose Absorbed by the Intestine

While this statement may seem to contradict the first one, both are correct. Metformin reduces glucose transport across the intestinal epithelia in the proximal jejunum as well as in the proximal ileum in a dose-dependent manner [73]. As all dietary glucose is almost completely absorbed in the upper small intestine, it is not possible to further increase the amount of glucose received by the enterocytes. Enterocytes need more glucose from the intestinal lumen or from the blood. If no food is taken, enterocytes themselves consume glucose from the blood.

In the blood that comes from the intestine to the liver there is a reduced amount of glucose and an increased amount of lactate. The concentration of this less energy-rich substance is increased up to 50% [66, 67]. The lactate concentration is significantly higher in the blood that is transmitted from the intestines to the liver than in the peripheral blood that we usually examine. A significant – 2.5-fold – increase in lactate concentration in the hepatic portal vein was recorded [74]. Under usual conditions, lactate is almost completely metabolized in the liver. In one study, metformin-treated patients had only slightly higher plasma lactate concentrations (in peripheral blood) than non-metformin-treated subjects, 1.86 vs 1.58 mmol/l respectively. Metformin-treated patients had a mean plasma lactate 0.16 mmol/l higher than subjects not taking the drug [75].

It should be remembered that the human body only gains from the nutrients in the colon what remains after the energy-rich substances have been drained by bacteria. Metformin also reduces the effectiveness of metabolism in bacteria residing in the colon. Its influence changes the intestinal microbiome. A common finding was a reduction in butyrate-producing bacteria and an increase in opportunistic pathogens [76-78]. Metformin cause an overall decrease in bacterial diversity [79, 80]. It can be expected that studies conducted in the future will demonstrate that the total amount of nutrients recovered in the colon is reduced by metformin.

14.3. .Metformin Absorption

Metformin concentrations in the jejunum peak at 500 μg/g, a value 30–300 times higher compared with its plasma concentrations [67]. After oral administration, metformin is slowly absorbed from the proximal small intestine (e.g., duodenum) [81, 82]. Metformin absorption is mediated by an active, saturable absorption process [81]. OCT1 and OCT3 play an important role in the transfer of metformin from the intestinal lumen to the interstitium. OCT3 is associated with metformin uptake and efflux in the salivary glands, which may account for the dysgeusia associated with metformin therapy [83].

Furthermore: Human thiamine transporter 2THTR-2 (hTHTR-2), SLC19A3, which is highly expressed in the small intestine, transports metformin too. The uptake mechanism for hTHTR-2 is pH- and electrochemical gradient-sensitive [50].

Metformin is efficiently taken up across the apical (luminal-facing) surface of enterocytes via bi-directional transporters, but that efflux across the basolateral surface of enterocytes is limited, resulting in metformin accumulation in the epithelium [84].

The human serotonin transporter (SERT, SLC6A4) plays a role in metformin intestinal absorption [47]. Serotonin (5-HT) release from the intestine is associated with nausea, vomiting and diarrhea, i.e., symptoms similar to those associated with metformin intolerance. Therefore, one possible mechanism for the GI intolerance associated with metformin use may relate to altered serotonin transport or to a direct serotonergic-like effect of metformin. Metformin uptake via SERT or OCT1 results in reduced serotonin transport and resultant GI side effects [85]

Metformin concentrations in the gastrointestinal (GI) lumen of patients taking 1-g oral doses are estimated to be 24 mM [50]. Thirty to 50% of patients on metformin report GI side effects such as diarrhea and bloating [86, 87].

Modified-release formulations of metformin have been developed to spread the absorption of metformin along the intestine thus reducing the drug's local concentrations, in an effort to raise its tolerability. A new metformin formulation has recently been developed, metformin XR (extended release) and metformin DR (delayed release), which is designed to target the ileum via pH-dependent dissolution of the tablet. The bioavailability of metformin DR is lower, yet its glucose-lowering efficacy is similar despite lower systemic metformin exposure [88].

What is the main effect of metformin in the intestine? This drug reduces the efficiency of (aerobic) metabolism and reduces the amount of glucose obtained from food.

15. The Liver

Patients with T2DM show increased rates of endogenous glucose production, which can be attributed to increased rates of gluconeogenesis. Metformin lowers the rate of glucose production in these patients through a reduction in gluconeogenesis.

The rate of gluconeogenesis is three times higher [94] in diabetic subjects compared with controls (0.59 vs. 0.18 mmol $\cdot$ $m^{-2} \cdot min^{-1}$). Metformin reduces that rate by 36% (to 0.38 mmol $\cdot$ $m^{-2} \cdot min^{-1}$) [88].

The above findings highlight the importance of correct interpretation of the phenomena observed in experiments. The liver of a patient with diabetes is not a perpetual motion machine. It is unable to create matter. Increased glucose production reflects increased entry of energy-rich substrates into the liver such as glucose, lactate, amino acids, free fatty acid and other energy-rich substances (e.g., so-called hepatic gluconeogenic substrate, i.e., alanine, glutamine, glutamate, glycerol, glycine, lactate, threonine or serine).

Even metformin is unable to annihilate the matter. The liver of metformin-treated patients receives smaller amounts of glucose precursors; moreover, its metabolism is less effective due to metformin use.

Nonetheless, experiments have produced quite different data. Hence the question arises which experiments provide accurate information about the effect of metformin on the liver? Experiments with low or high metformin concentrations? The answer is: both. After oral administration, the plasma concentrations of metformin peak within 1 hour in man as well as in animals [70, 88, 89] to achieve values of 10–40 μM in humans following a therapeutic dose [70, 89, 90]. As metformin is absorbed from the GI tract and delivered directly to the liver through the portal vein, the plasma concentrations of metformin in the portal vein are relatively high (40–80 μM) [70]. The most accurate results will be obtained by experimenting on cell cultures with this (80 μM) concentration for several days. However, in order to define the influence of all confounders and regulatory mechanisms, the duration of trial should be months or years.

There was in other studies considerable glycogen cycling in individuals with poor control of diabetes, which accounted for about 25% of endogenous glucose production. Hyperglucagonemia and hyper-insulinemia, conditions typically present in T2DM patients, have both been shown to promote glycogen cycling [91-93]. However, glycogen stores in diabetic patients, whether treated or not with metformin, are not extremely elevated or reduced in the long term. Thus, this mechanism is of little significance and only temporary.

Metformin treatment also lowered the plasma free fatty acid concentrations by 30% in the diabetic subjects [94-96]. Reduction in plasma free fatty acid concentrations after metformin treatment also contributed to reduced rates of gluconeogenesis [88].

Metformin decreases the rate of hepatic lactate uptake while increasing the amount of lactate taken up by the liver. Which of the two claims is correct? Both.

Metformin was observed to decrease gluconeogenesis in perfused liver, primarily through inhibition of hepatic lactate uptake [105]. The drug would have reduced lactate entry to liver cells if the lactate levels were the same as in those not treated with metformin. However, since lactate production (in the intestine or in the muscles) in metformin-treated patients is increased and the lactate levels in the portal blood are increased, the liver's lactate entry is not reduced but elevated.

Metformin decreases hepatic glucose production [67, 88, 98, 99]. Restricted energy income, restricted ATP production and increasing AMP concentrations influence the basic mechanism defining energy gain of the cells. Reduced amounts of some energy-rich substrates and lack of energy (ATP) in hepatocytes curb the ability of the liver to produce glucose and thus reduce total glucose output from the liver to the blood. The total effect is given by the sum of the simple ATP deficiency and the influence of the regulatory mechanisms (LKB1, AMPK).

Metformin inhibits gluconeogenesis from L-lactate in isolated hepatocytes in a time- and concentration-dependent fashion. This is consistent with the time-dependent accumulation of the drug within the mitochondria and subsequent inhibition of the respiratory chain. After 2.5 hours of incubation with 2 mM metformin, the percentage of inhibition reached 75% [5].

The K0.5 (50% efficacy) value with respect to extra-mitochondrial metformin is about 80 μM at equilibrium and is consistent with the 29% inhibition of ADP-stimulated glutamate-malate oxidation observed in the mitochondria from cultured rat hepatoma cells exposed to 50 μM metformin for 60 hours. Fifty percent inhibition of respiration led to 20% inhibition of gluconeogenesis [100].

Metformin, via an AMP-activated protein kinase (AMPK)-dependent mechanism, suppresses glucose production and gluconeogenic gene expression in primary hepatocytes at concentrations found in the portal vein of (60–80 μm). Oral metformin inhibits gluconeogenic gene expression in the liver. Furthermore, the cAMP-PKA pathway negatively regulates AMPK activity through phosphorylation at Ser-485/497 on the α subunit, which in turn reduces net phosphorylation at Thr-172. Because diabetic patients often have hyperglucagonemia, AMPKα phosphorylation at Ser-485/497 is a potential target of metformin efficacy [101].

What is the main effect of metformin in the liver? The drug reduces the efficiency of metabolism and reduces the ability to produce glucose.

16. Muscle

Muscles make up a significant part of human body weight. Optimally, 35% (female) to 40% (male) of a healthy person's body mass is made up by skeletal muscle. Any change in muscle metabolism will entail a major change in the need for glucose and fatty acids, and significantly change blood glucose levels. Metabolism is predominantly aerobic. Especially the slow twitch (red) fibers focused on long-term or repetitive contraction are powered by aerobic processes.

Muscles that are capable of rapid physical performance work on the principle of anaerobic glycolysis. Fast twitch (white) fibers contain lesser amounts of mitochondria. Anaerobic metabolism allows faster development of muscle strength but its by-product is lactic acid converted back to glucose in the liver.

Most of the processes that continually consume ATP energy (e.g., pumps that transfer ions against concentration gradients) are dependent on the aerobic (more energy-efficient) process (in both red and white muscle).

Insulin increases glucose input into the muscle. However, when a certain upper limit is exceeded, further glucose input into the cell becomes unsuitable, and the internal regulatory mechanisms begin to limit glucose supply into the muscle. A secondary feedback event arises, which is referred to as "resistance". Maintenance of high insulin and glucose blood concentrations will trigger feedback the regulatory mechanism fast, within seven days [102].

Insulin increases glycogen production in the muscle. While the amount of glycogen is slightly increased, it does not reach the levels found that patients with an inherited glycogen-type metabolic disorder. A regulatory mechanism that detects excessive amounts of glycogen produces counter-regulatory events that curbing further glycogen production.

(Compare: glycogen storage diseases: glycogen synthase, acid alpha glucosidase, glycogen debranching enzyme, glycogen branching enzyme, muscle glycogen phosphorylase, muscle phospho-fructokinase, phosphoglycerate mutase, aldolase A, beta-enolase, glycogenin 1 and other deficiencies: Pompe's, Cori's, Forbes', Andersen, Mc Ardle diseases.)

Insulin increases the production of lipids in the muscle. The patient suffering from diabetes (T2DM) has intramyocellular lipid stores comparable to those of endurance athletes [103], but will not use them.

Metformin inhibits aerobic metabolism even in red and resting white muscle. The drug reduces the rate of ATP production and, conversely, causes increased concentrations of ADP or AMP. The end product of glucose metabolism is not CO_2 (with 32 ATP molecules obtained) but, to some extent, lactate (with a 2 ATP molecule gain). The need for glucose in the muscle increases. Lactate enters the Cori cycle: it is transmitted from the site of origin (the muscle) to where glucose is produced in (the liver), but at the cost of losing 4 ATP molecules to recover each glucose molecule. (The efficiency of this process is about 33%, 11 ATP/glucose.)

Likewise, metformin increases skeletal or heart myocyte glucose uptake [104-106], directly via AMP, AMP/ATP, via AMPK (AMPKα2) [107], via pyruvate dehydrogenase inhibition [52], via mitochondrial glycerol-phosphate dehydrogenase inhibition [61], or via fructose-1-6-bisphosphatase inhibition [63]. What is the main effect of metformin in muscle? This drug reduces the efficiency of (aerobic) metabolism and increases the amount of glucose consumed by the muscle.

17. Adipose Tissue

In many people, the mass of adipose tissue significantly exceeds that of muscle tissue. Adipose tissue is highly metabolically and hormonally active. Adipose tissue secretes hormones that contribute to energy homeostasis (adipokines such as leptin, adiponectin, adipsin, interleukin-6, tumor necrosis factor-α and others). Leptin acts through the brain to regulate food intake and energy expenditure [109]. Another type of adipose tissue serves as an energy storage depot to maintain lipid homeostasis.

Insulin increases fat synthesis and fatty acid esterification while decreasing lipolysis. Disorders caused by severe insulin resistance (Rabson-Mendenhall syndrome, Donohue syndrome) are characterized by a deficiency or absence of adipose tissue.

Metformin treatment brought in a study significant reduction in visceral fat mass compared to controls accompanied by up-regulation of a fat oxidation-related enzyme in the liver, UCP-1, in brown adipose tissue [110]. Besides· the drug was shown to reduce the amount of body fat and improve body composition in studies performed in patients with T2DM [96, 111]. It has also been shown to enhance the expression of uncoupling proteins (UCPs), the key enzymes of adaptive thermogenesis [112]. Thus, it has been hypothesized that metformin administration promotes the reduction of body fat amount via acceleration of fat oxidation and adaptive thermogenesis *in vivo*. Has increased production of lactic acid been proven in adipose tissue due to the metformin therapy? Yes. The rate of net postabsorptive subcutaneous lactate release was increased (from 149 to 475 nmol $\cdot$ 100 g^{-1} $\cdot$ min^{-1}) [113]. What is the main effect of metformin in adipose tissue? The drug reduces the ability to store fat deposits.

18. Brain

Metformin may also exert an anorectic effect as metformin-treated patients tend to receive smaller amounts of food.

Besides, metformin has a direct anorexigenic effect [114-117]. It increases the plasma levels of the anorectic gut hormone peptide YY (PYY) via cell internalization through PMAT and SERT and intracellular activation of AMPK [188]. Other effects are also most likely – transporter interactions in the gastrointestinal tract, signs of local thiamine deficiency [55-57] and others. It has been hypothesized that metformin activates GLP-1 receptors to increase protein kinase A (PKA) activity on intestinal vagal afferents (gut-brain axes) [119].

The effect of metformin on the brain (through what mechanism this drug reduces the craving for food) is also believed to be extremely important, but evaluation of any possible effects warrants further long-term research.

19. Side Effects

Generally, metformin is safe, with the most frequent adverse effects being gastrointestinal disorders, nausea, vomiting, flatulence and diarrhea. The most serious side effect is lactic acidosis, i.e., metformin-induced lactic acidosis (MILA). Metformin is contraindicated in patients developing hypoxic states and severe renal impairment (effective glomerular filtration, eGFR below 30 ml/min/1.73 m^2)!

A systematic review concluded no data exists to definitively link metformin to lactic acidosis [120]. Lactic acidosis never occurs with metformin exposure during routine medical care [121]. The incidence of MILA is about 9 per 100,000 person-years [122].

Unfortunately, we cannot confirm these optimistic data. The incidence of MILA in our catchment area of about 90,000 inhabitants is relatively high. Overall, 27 cases of MILA were identified over a period of 21 years (1998–2019), i.e., 21.3 cases a year per 100 000 patient-years. Mixed disturbances of acid-base balance (with other acidifying but, also, alkalizing disorders) were usually present. Blood pH was on average low (pH 7.1), with the lowest and highest values being 6.65 and 7.47, respectively. The patients usually presented with significant hypocapnia (average pCO_2, 2.73 kPa). Hypochloremia caused alkalizing disorder. Lactate levels were 11.9 mmol/l (max. 19.8 mmol/l). Levels of Base Excess, BE were on average lowered (by 20.2 mmol/l) as were bicarbonate levels (9.06 mmol/l). Levels of Anion Gap, AG were high: 37.97 mmol/l. Patients' partial pressure of oxygen was relatively high (pO_2, 13.2 kPa). Patients were usually diagnosed to have pre-renal acute failure (creatinine level 575 umol/l). Renal function gradually became fully normalized in most survivors. Levels of glycemia were on average high (14.6 mmol/l), with hypoglycemia detected in only five cases. The highest plasma metformin concentration was 61.1 mg/l (i.e., 473 μmol/l).

Hyperventilation and hypocapnia seem to be the underlying mechanisms preventing the development of critical intracellular acidosis in MILA.

The basic sign of metformin intoxication was organ failure, especially of those organs that require a lot of energy. Disorientation or unconsciousness was indicative of impaired cognitive function. Patients who were able to speak reported severe vomiting and diarrhea.

The first significant disturbance and, probably, also the cause of MILA development in our patients was gastrointestinal tract involvement. While common infection – or the effect of metformin itself – caused vomiting and diarrhea, subsequent hypovolemia resulted in pre-renal failure, which in turn led to an increase in metformin levels up to toxic values. Development of kidney failure was often unexpected and fast. The most effective therapeutic option in our group of patients was elimination of metformin from the body (using hemodialysis or continuous veno-venous hemodialysis, CVVHD).

Metformin is generally a safe drug, significantly safer than phenformin, buformin (and probably relatively safer than insulin). However, we believe that MILA is a serious condition with an incidence significantly higher than reported in the relevant literature.

20. Insulin

Insulin is one of the most important anabolic hormones.

Insulin stimulates the translocation of the GLUT4 from intracellular pools to the surface cell membrane [123]. Besides, the hormone stimulates glycogen synthase, the key regulating enzyme for glycogen synthesis, and many other enzymes.

Let us mention its main "clinical" effects:

Lipid metabolism:

1. Insulin stimulates triacylglycerol synthesis in adipose tissue, muscles and the liver
2. Insulin increases the rate of cholesterol synthesis in the liver
3. Insulin decreases the rate of lipolysis in adipose tissue
4. Insulin increases the uptake of triglycerides from the blood into adipose tissue and muscles
5. Insulin decreases plasma fatty acid levels

Protein metabolism:

6. Insulin increases the rate of protein synthesis in muscles, the liver and other tissues
7. Insulin decreases the rate of protein degradation in muscles
8. Insulin increases the rate of transport of amino acids into tissues

Minerals:

9. Insulin decreases renal sodium excretion
10. Insulin promotes the transfer of potassium to cells

Carbohydrate metabolism:

11. Insulin inhibits the rate of gluconeogenesis in the liver
12. Insulin increases the rate of glycogen synthesis in muscles, adipose tissue and the liver
13. Insulin decreases the rate of glycogen breakdown in muscles and the liver
14. Insulin enables glucose to enter into insulin-sensitive tissues (muscles, adipose tissue, and other cells containing the GLUT4 transporter)
15. Insulin decreases the blood glucose concentration

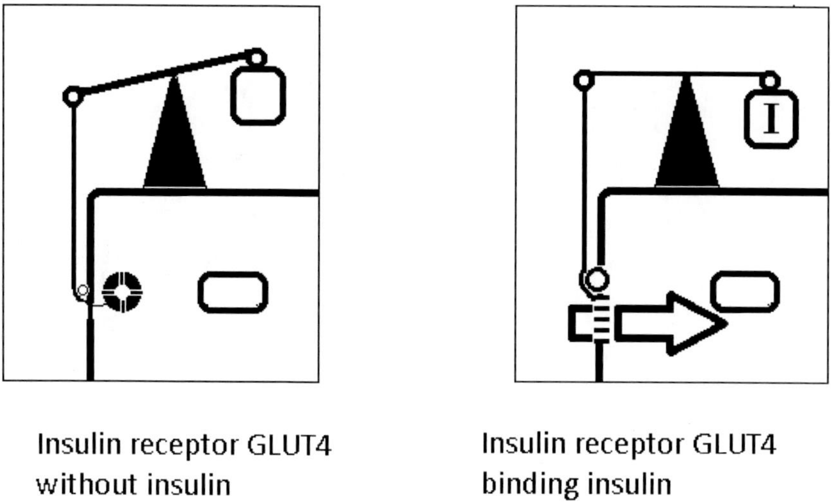

Figure 5. GLUT 4. Insulin "opens" the "cellular gate" via GLUT4 in the muscle. This reaction enables glucose transport from blood into the cell. [129, 135].

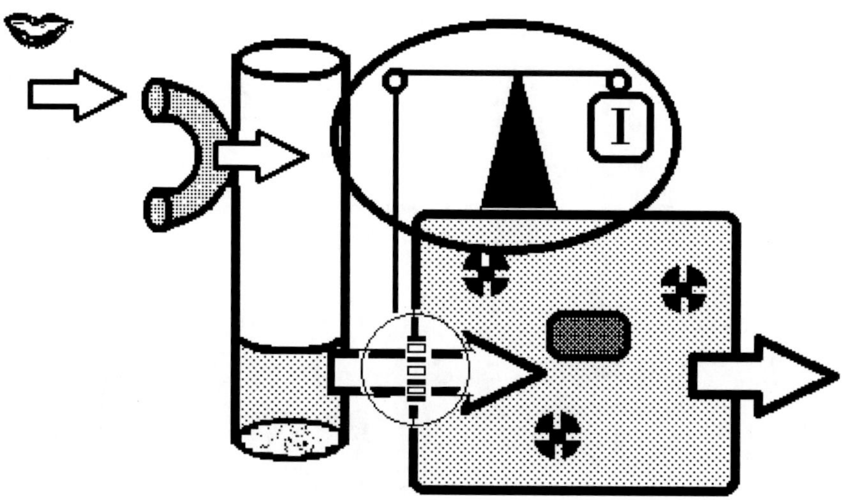

Figure 6. Insulin gate. Insulin "opens" the "cellular gate" via GLUT4 in the muscle. This reaction enables glucose transport from blood into the cell [129, 135].

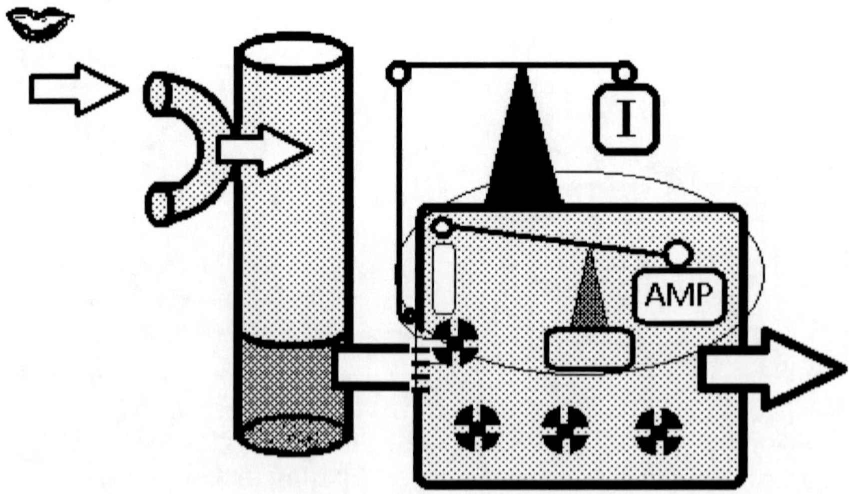

Figure 7. Intracellular gate. AMPK and other members of the AMPK family determines the amount of glucose entering the cell [129, 135].

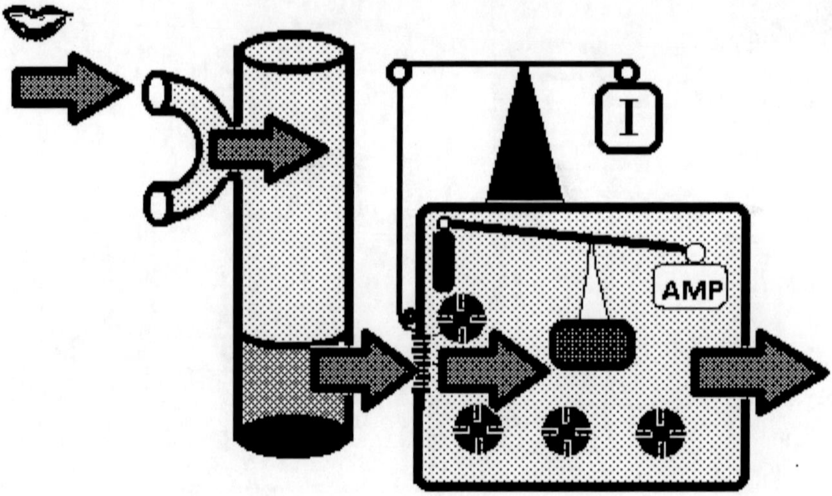

Figure 8. Physiological situation. Food intake and energy expenditure are in the optimal correlation [129, 135].

A sudden increase in insulin concentration – and, hence, a rapid surge in its effect – causes a rapid fall in blood glucose levels (point 15).

A slow yet steady increase in insulin concentrations and, hence, chronically enhanced effect of insulin primarily promotes the anabolic processes (items 1 to 14). The blood glucose levels depend on the rate of delivery of energy-rich substances to the body, which is usually normal or elevated. The patients du not suffer from *lack of insulin* but they suffer from *surplus of insulin* (and from transformation of its effect). While the term "*hyperinsulinism*" is rarely used today, it most befittingly reflects the real-life situation. Hyperinsulinism and *insulin uncoupling*:

21. Insulin Resistance and Insulin-Uncoupling (Pseudo-Resistance)

Insulin resistance (IR) is a condition whereby cells fail to respond to the actions of insulin with the result being that insulin-resistant cells cannot take in glucose.

Primary *congenital* insulin resistance is a rare disorder caused by the lack of a fully functional insulin receptor, a condition also referred to as Donohue syndrome [124, 125] or Rabson–Mendenhall syndrome [126]. None of the clinical effects of insulin (items 1–15) is expressed.

By contrast, *acquired* insulin resistance is a more frequent phenomenon occurrence. Severe infection, sepsis, burns and other stressful conditions impair glucose metabolism and involve insulin resistance. In these cases, the muscle is rapidly degraded. The released amino acids are converted to glucose. Catabolic processes are harmful for the whole body of the patient.

Quite a different situation occurs in the case of "insulin pseudo-resistance", "resistance" present in obese diabetics.

This phenomenon arose from misinterpretation of experiments with the induction of "comparable conditions". In clamp studies, we ensure completely identical conditions for diabetics and healthy control persons. Insulin and glucose levels are identical [127, 128]. Under these conditions, paradoxically, there is less glucose input into the obese diabetic tissues than in controls. What is the explanation for this phenomenon?

In fact, we have mixed up the manifestations of the *disorder* and the *feedback* mechanism that protects the diabetic cells from overfilling with energy-rich substrates.

As the disease advances, the excessive glucose uptake by adipose tissue cells and muscle is ensured by the high concentration gradient, hyperglycemia and hyperinsulinemia. Induction of "comparable conditions" in clamp studies leads to paradoxical results. Due to relative hypoglycemia and hypoinsulinemia (as compared to normal conditions) – during the clamp experiment – the tissues of diabetic patients take up a smaller amount of glucose than those of non-diabetic subjects. This phenomenon is called the "paradox of insulin resistance" [129]. Reduced glucose input to cells takes only a short time while we perform the experiment and is a manifestation of intracellular feedback.

On the contrary, glucose uptake by the muscle and adipose tissue are *higher* in obese diabetics during their previous as well as subsequent life.

Insulin pseudo-resistance can be induced by mere maintenance of hyperinsulinemia [102]. It can be minimized by reducing nutrient intake and by increasing physical exertion [130-132]. The clinical effects of insulin (items 1–14) are fully expressed. Only the last item – item 15 – is not accomplished, with hyperglycemia persisting in patients. This is, however, not due to an impaired insulin effect, but to an excessive energy intake.

In the text below, we use the term "insulin uncoupling", a term expressing disconnection (split) of insulin functions.

22. Unusual Comparison

The following lines present a seemingly absolutely unscientific, but well-remembered comparison.

Let's compare the human body with a well-functioning car.

This car would be able to drive at 200 kilometers per hour for 14 hours a day. The car would travel 2800 kilometers a day and its fuel consumption would be 1 liter of gasoline per hundred kilometers, thus 28 liters per day.

Imagine that we would like to use 50 liters of gasoline (much more than needed) every day and drive just 500 kilometers with this car. And every other day we would have to draw another 50 liters into the car.

Nobody can imagine that we would want a worse car that needs more fuel. Such a wish is absolutely pointless. Yet the contemporary man wants their body to consume a large amount of food without moving. The question is: *How do we have to damage the car to meet the above owner's requirements?*

One option is to make a small hole in the tank to permanently lose fuel. This corresponds to the effect of gliflozin. This drug allows urine glucose losses.

Another option is to pour gasoline inside the car. Unfortunately, there is no room left for passengers, but more gasoline can be poured. This is how (high doses of) insulin and sulfonylureas work. High doses of insulin (high insulin concentrations) makes it possible to store energy-rich substances in cells. The most preferred form of energy storage is fat. Excess of insulin causes creates fat deposits in the organs. The side effects include morbid obesity, liver steatosis or myocardial steatosis.

Another possibility is to convince the gas station attendant not to pump 50 liters but less. Anorectics work on this principle. Above all, incretins reduce our desire for food. Also metformin reduces our craving for food.

The last option is to pull the handbrake. The car is able to drive but consumes more gasoline. This will allow you to draw more fuel as we wish. Metformin ensures exerts this effect in a safe and predictable way.

Preparations that acted too strongly (phenformin and buformin) caused complete loss of metabolic processes and are currently banned. Glitazones also reduce the effectiveness of metabolism too intensely so their use is limited today.

In fact, the drugs we are now more or less successfully treating T2DM are actually causing the loss of energy-rich substances.

23. Diagnostic Entities within the Chronic Hyperglycemic Syndrome

Regarding cell energy intake, there are two main groups of diagnostic entities:

23.1. Diabetic Syndrome with Intracellular Starvation

This type includes the following diagnoses:

- True type-1 diabetes mellitus (T1DM), i.e., diabetes mellitus that occurs in childhood. This disease is characterized by real lack of insulin. More precisely: Untreated T1DM is characterized by lack of insulin and intracellular starvation.
  Diabetes mellitus results from autoimmune destruction of the insulin-producing beta cells in the pancreas and subsequent lack of insulin.
- Latent autoimmune diabetes in adults (LADA) is the identical form of T1DM, a form that occurs in adulthood. It is characterized by autoimmune destruction of the insulin- producing cells and, also, by real lack of insulin.
- Diabetes mellitus with true insulin resistance. Insulin resistance (IR) is a condition whereby cells fail to respond to the actions of insulin. Insulin-resistant cells cannot take in glucose. Though extremely rare, this disorder does occur and is termed Donohue syndrome or leprechaunism [134, 135] caused by a mutation of the insulin receptor [143, 144]. The mutant allele is not as effective as it should be, the usual doses of insulin are insufficient and the insulin-dependent cells begin to starve.
- Others diseases with real lack of insulin. For example: diseases caused by the destruction or removal of pancreatic beta-cells (i.e., after surgical pancreatectomy due to injury or carcinoma), intoxications et al.

- A number of other diseases that are associated with genetic disorders at the level of individual enzymes within intracellular glucose processing, intra-mitochondrial ATP production, or many diseases associated with other hormone and regulatory disorders.

23.2. Diabetic Syndrome with Intracellular Overfeeding

- Diabetes mellitus type 2 (T2DM) with obesity. This metabolic disorder is not caused by a relative (or absolute) lack of insulin, nor is it caused by inborn insulin resistance. [135]. T2DM is caused by insulin uncoupling.

Type-2 diabetes mellitus is a disease caused by chronic excessive energy intake that has damaged the basic regulatory systems [135]. Other contributing factors include excessive intake of energy-rich substances [102], excess of food, and lack of physical exercise. Type-2 diabetes mellitus with obesity is characterized by excessive filling of cells by energy-rich substances. The patients create excessive intra-myocellular fat deposits comparable to those of endurance athletes [138]. In fact, T2DM develops due to a collision of basic (extracellular) hormonal systems (insulin and other hormones) and the intracellular regulatory system (AMPK and other "intracellular hormones") [135]. T2DM is caused by inadequate, unhealthy lifestyle - incompatible with the genetic predisposition of the overwhelming proportion of the general population [136, 139, 140], a population with a thrifty genotype [137]. Diabetes with obesity is the result of maladaptive habits, amenable to behavioral therapy.

23.3. Combinations

The disorders that afflict the aforementioned patient populations can combine. As the number of patients suffering from diabetes is extremely high, it is well possible that the patient develops one or more of the congenital as well as acquired types of the hyperglycemic syndrome.

The following two disorders are common: T1DM causing T2DM and T2DM causing T1DM!

Many patients treated for T1DM (syndrome of T1DM with intracellular starvation due to lack of insulin) tend to overfeed and, consequently, become hyperglycemic. As a result, they require high, supra-normal, doses of insulin. These patients, of necessity, suffer from intracellular overfeeding – and obesity – and, hence, inevitably develop T2DM.

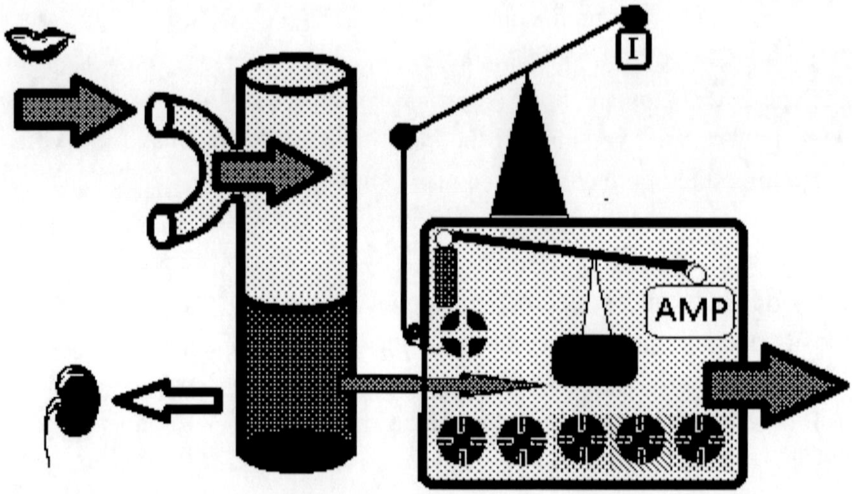

Figure 9. T1DM. T1DM results from the destruction of the insulin-producing beta cells and the subsequent lack of insulin. Insulin "gate" is closed. The most important symptoms are intracellular starvation, shortage of intracellular ATP [135].

On the other hand, many obese patients with T2DM receiving treatment with sulfonylureas gradually lose their ability of insulin production (or they suffer from LADA) to eventually develop real, absolute lack of insulin (typical of T1DM). They suffer from both type of diabetes mellitus T2DM and T1DM. In these cases, the combination of metformin and insulin will be beneficial.

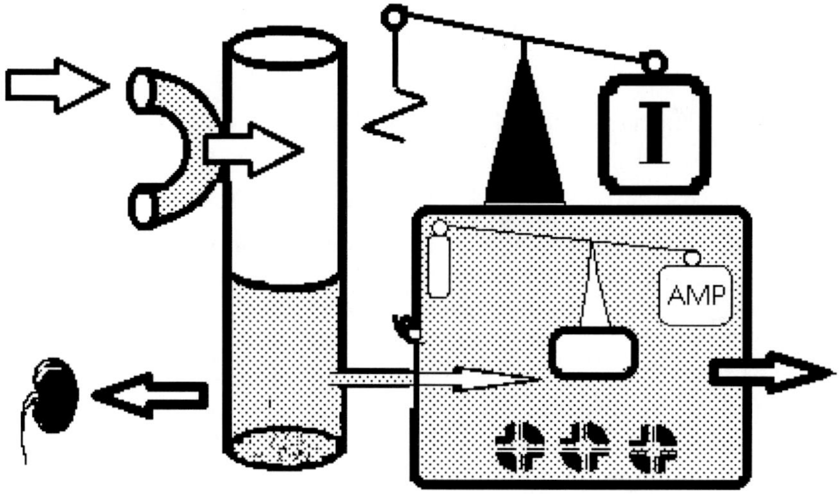

Figure 10. Donohue syndrome. Insulin resistance (IR) is a condition in which cells fail to respond to the actions of the insulin. Insulin resistant (GLUT4-dependent) cells cannot take in glucose. Subcutaneous adipose tissue and muscle are diminished [135].

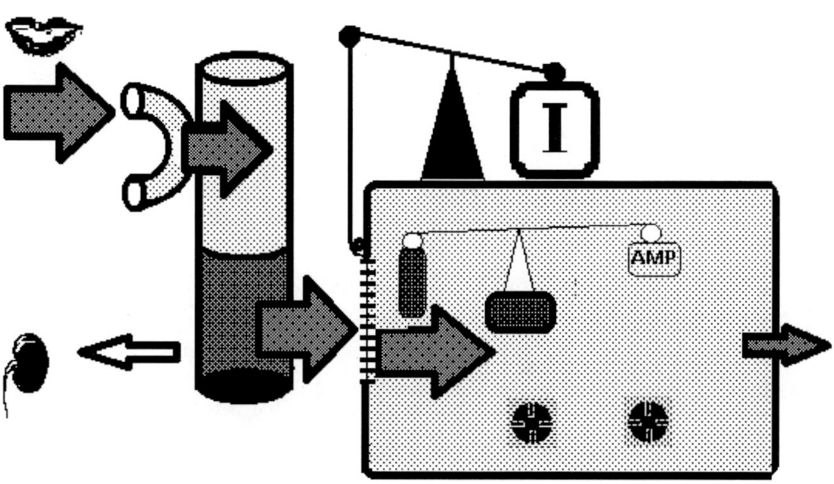

Figure 11. Diabetes mellitus type 2, T2DM. The Insulin production is high, (but it gradually decreases). The input of the glucose into cells is increased. Energy expenditure is low. The blood glucose level is high and rising. There are no signs of intracellular starvation. The insulin uncoupling ("pseudo-resistance") is present [135].

24. Other Diseases

24.1. Cardiovascular Effect of Metformin

The prognosis of patients with T2DM used to be grim. Despite being treated with sulfonylureas or insulin, they would die significantly earlier from cardiovascular and other diseases. The UK Prospective Diabetes Study (UKPDS) and a number of other projects (which were subsequently performed) showed that the effect of metformin is significantly more beneficial than that of sulfonylureas. Why? Probably because metformin reduces hyperglycemia but has no other (insulin-like) anabolic effects. The UKPDS investigators reported that metformin treatment lowered the risk of myocardial infarction (by 39%) when compared with traditional treatments over a period of 10 years [141] as well as during subsequent 10-year follow-up [142].

24.2. Non-Alcoholic Fatty Liver Disease

With a prevalence of approximately 25% worldwide, non-alcoholic fatty liver disease (NAFLD) is thought to be currently the most common liver disease in the world [143]. Metformin was considered to be very usefull in treatement of the NAFLD - and the first information from a meta-analysis indicated that metformin improves liver function, ALT, AST (but not histological response) in NAFLD patients [144]. Regrettably, no studies have shown whether metformin improves long-term outcomes. It is to be hoped that experiments carried out in the future will offer more favorable and robust data.

24.3. Obesity

Metformin has been shown to have beneficial effects in the treatment of obesity [145-148].

Let us present the results of several studies.

24.3.1. Effectiveness of Metformin on Weight Loss in Non-Diabetic Individuals with Obesity

In one study, treatment with metformin for 6 months was successful, with mean weight loss in the metformin-treated group being 5.8 kg (5.6%) whereas untreated controls gained 0.8 kg (0.8%) on average. Patients with severe insulin resistance lost significantly more weight as compared to insulin-sensitive patients. The percentage of weight loss was independent of age, sex or BMI [145]. In another report, metformin reduced, compared with placebo, BMI by 1.42 kg/m^2 in obese children and adolescents [146]. Yet other authors showed that six months of metformin therapy improved weight loss and reduced abdominal adiposity in obese children with normal glucose tolerance. Weight loss was modest but more pronounced in the metformin group (-4.9 kg) than in the diet/exercise group (-1.7 kg) [147]. Similarly· metformin treatment significantly reduced body weight after 1-month hypocaloric dieting and 6-month combined treatment with a hypocaloric diet plus metformin. Another study in women with preexistent abdominal or visceral obesity and polycystic ovary syndrome reported a weight loss of 3 kg versus 7 kg in abdominally obese women not developing polycystic ovary syndrome. Visceral adipose tissue area values significantly decreased during metformin treatment [148]. These results are most encouraging.

24.4. Cancer

Today, metformin is generally viewed as a promising candidate drug for cancer treatment in combination with other cytotoxic agents. Many clinical trials designed to investigate metformin anticancer activity including breast cancer, lung, esophagus, liver, pancreatic cancer or leukemia are currently underway (see overview) [149 et al.].

25. Metformin – For Whom

Metformin was discovered in 1922 to be subsequently forgotten for many years although its blood glucose-lowering effect was known. Jean Sterne reported the use of metformin to treat diabetes in 1957. The long-term cardiovascular benefits of metformin were identified by the UK Prospective Diabetes Study (UKPDS) in 1998. Metformin has become the preferred first-line oral blood glucose-lowering agent to manage type 2 diabetes mellitus [150].

Strictly speaking – we use this drug for the treatment of "syndrome of intracellular overfeeding - with secondary syndrome of chronic hyperglycemia", today known as "syndrome T2DM", or "Type-2 diabetes mellitus with obesity".

At present, metformin is the drug of choice for the treatment of T2DM in obese individuals. While not recommended in cases of DM syndrome with intracellular starvation, i.e., real T1DM (a disease which is characterized by lack of insulin), the agent can be useful in patients with *either* type of diabetes mellitus, i.e., both T1DM and T2DM.

Metformin has also been shown to have beneficial effects in the treatment of obesity. When it becomes one of the basic drugs used in the management of this serious disease is a difficult question to answer. After thirty years (as in the case of diabetes treatment)?

Today, metformin belongs to the safest medications in its class. Nevertheless, if forced to use drugs that limit the influence of energy surplus and decrease the effect of metabolism, we can still expect an adverse effect characterized by aerobic metabolism lapse, MILA. Therefore, we do not use high doses of metformin alone but a combination of metformin with other drugs such as incretins, gliflozins or even insulin. While adequate food intake and physical training have the same beneficial effect, their advantage no one can deny is that they do not cause any harm.

25.1. Summary

The main effect of metformin is apparently *a decrease in metabolism efficiency*. It is a medicine that eliminates the problem of *damage caused by energy excess*. So what is the likely number of patients for whom metformin would be beneficial? One or two billion today?

References

[1] Kubat, K; Zboril, M. Undesirable effects of biguanide: 31 cases of lactic acidosis in patients treated with buformin in the course of 10 years. *Klin. Biochem. metab.*, 2000, 8(29), 103-107.

[2] Kubat, K; Zboril, M; Semradova, M; Kanak, V. 23 cases of metformin-induced metabolic lactic acidosis in patients treated with metformin. *Klin. Biochem. Metab.*, 2017, 25(46), No 2., 77-85.

[3] Lalau, JD; Lemaire-Hurtel, AS; Lacroix, C. Establishment of a database of metformin plasma concentrations and erythrocyte levels in normal and emergency situations. *Clin. Drug. Investig.*, 2011, 31(6), 435-438.

[4] Brunmair, B; Staniek, K; Gras, F; Scharf, N; Althaym, A; Clara, R; Roden, M; Gnaiger, E; Nohl, H; Waldhausl, W; et al. Thiazolidinediones, like metformin, inhibit respiratory complex I: a common mechanism contributing to their antidiabetic actions? *Diabetes.*, 2004, 53, 1052–1059.

[5] Owen, MR; Doran, E; Halestrap, AP. Evidence that metformin exerts its anti-diabetic effects through inhibition of complex 1 of the mitochondrial respiratory chain. *Biochem. J.*, 2000, 348, 607–614.

[6] Chappell, JB. The effect of alkylguanidines on mitochondrial metabolism. *J. Biol. Chem.*, 1963, 238, 410–417.

[7] Davidoff, F. Effects of guanidine derivatives on mitochondrial function. III. The mechanism of phenylethylbiguanide accumulation and its relationship to *in vitro* respiratory inhibition. *J. Biol. Chem.*, 1971, 246, 4017–4027.

[8] El-Mir, MY; Nogueira, V; Fontaine, E; Averet, N; Rigoulet, M; Leverve, XJ. Dimethylbiguanide inhibits cell respiration via an indirect effect targeted on the respiratory chain complex I: *J Biol. Chem.*, 2000, 275, 223–228.

[9] Larsen, S; Rabøl, R; Hansen, CN; Madsbad, S; Helge, JW; Dela, F. Metformin-treated patients with type 2 diabetes have normal mitochondrial complex I respiration. *Diabetologia*, 2012, 55, 443–449.

[10] Zhang, L; He, H; Balschi, JA. Metformin and phenformin activate AMP-activated protein kinase in the heart by increasing cytosolic AMP concentration. *Am. J. Physiol. Heart Circ. Physiol.*, 2007, 293, H457–H466 10.1152/ajpheart.00002.2007.

[11] Xiao, B; et al. Structure of mammalian AMPK and its regulation by ADP. *Nature*, 2011, 472, 230–233.

[12] Oakhill, JS; et al. AMPK is a direct adenylate charge-regulated protein kinase. *Science*, 2011, 332, 1433–1435.

[13] Hawley, SA; Davison, M; Woods, A; Davies, SP; Beri, RK; Carling, D; Hardie, DG. (November 1996). "Characterization of the AMP-activated protein kinase kinase from rat liver and identification of threonine 172 as the major site at which it phosphorylates AMP-activated protein kinase". *J. Biol. Chem.*, 271 (44), 27879–87.

[14] Xie, Z; Dong, Y; Scholz, R; Neumann, D; Zou, MH. Phosphorylation of LKB1 at serine 428 by protein kinase C-zeta is required for metformin-enhanced activation of the AMP-activated protein kinase in endothelial cells. *Circulation*, 2008, 117, 952-962.

[15] Hindi, SM; Sato, S; Xiong, G; et al. TAK1 regulates skeletal muscle mass and mitochondrial function. *JCI Insight.*, 2018, 3(3), e98441. doi: 10.1172/jci.insight.98441.

[16] Hawley, SA; et al. Calmodulin-dependent protein kinase kinase-beta is an alternative upstream kinase for AMP-activated protein kinase. *Cell Metab.*, 2005, 2, 9–19.

[17] Hurley, RL; et al. The Ca2+/calmodulin-dependent protein kinase kinases are AMP-activated protein kinase kinases. *J Biol. Chem.*, 2005, 280, 29060–29066.

[18] Woods, A; et al. C(Ca2+)/calmodulin-dependent protein kinase kinase-beta acts upstream of AMP-activated protein kinase in mammalian cells. *Cell Metab.*, 2005, 2, 21–33.

[19] Fogarty, S; et al. Calmodulin-dependent protein kinase kinase-beta activates AMPK without forming a stable complex: synergistic effects of Ca2+ and AMP. *Biochem J.*, 2010, 426, 109–118.

[20] Jenne, DE; Reimann, H; Nezu, J; Friedel, W; Loff, S; Jeschke, R; Müller, O; Back, W; Zimmer, M. Peutz-Jeghers syndrome is caused by mutations in a novel serine threonine kinase. *Nature Genetics.*, 1998, 18(1), 38–43.

[21] Lizcano, JM; Göransson, O; Toth, R; Deak, M; Morrice, NA; Boudeau, J; Alessi, DR. LKB1 is a master kinase that activates 13 kinases of the AMPK subfamily, including MARK/PAR-1. *The EMBO journal*, 2004, 23(4), 833–843.

[22] Hayashi, T; Hirshman; MF; Kurth; EJ; Winder; WW; Goodyear; LJ. Evidence for 5′ AMP-activated protein kinase mediation of the effect of muscle contraction on glucose transport. *Diabetes*, 1998, 47, 1369-1373.

[23] Merrill; GF; Kurth; EJ; Hardie; DG; Winder; WW. AICA riboside increases AMP-activated protein kinase, fatty acid oxidation, and glucose uptake in rat muscle. *Am J Physiol*, 1997, 273, E1107-E1112.

[24] Goodyear, LJ. AMP-activated protein kinase: a critical signaling intermediary for exercise-stimulated glucose transport? *Exerc Sport Sci Rev*, 2000, 28, 113-116.

[25] Hardie, DG; Carling, D. The AMP-activated protein kinase: fuel gauge of the mammalian cell? *Eur J Biochem*, 1997, 246, 259-273.

[26] Bolster, DR; Crozier, SJ; Kimball, SR; Jefferson, LS. AMP-activated protein kinase suppresses protein synthesis in rat skeletal muscle through down-regulated mammalian target of rapamycin (mTOR) signaling. *Biol Chem.*, 2002 Jul 5, 277(27), 23977-23980.

[27] Krawiec, BJ; Nystrom, GJ; Frost, RA; Jefferson, LS; Lang, CH. AMP-activated protein kinase agonists increase mRNA content of the

muscle-specific ubiquitin ligases MAFbx and MuRF1 in C2C12 cells. *Am J Physiol Endocrinol Metab.*, 2007, 292(6), E1555–E1567.

[28] Sakamoto, K; McCarthy, A; Smith, D; Green, KA; Grahame, Hardie, D; Ashworth, A; Alessi, DR. Deficiency of LKB1 in skeletal muscle prevents AMPK activation and glucose uptake during contraction. *EMBO J.*, 2005 May 18, 24(10), 1810-1820.

[29] Suter, M; Riek, U; Tuerk, R; Schlattner, U; Wallimann, T; Neumann, D. Dissecting the role of 5′-AMP for allosteric stimulation, activation, and deactivation of AMP-activated protein kinase. *J Biol Chem*, 2006, 281, 32207–32216.

[30] Wojtaszewski, JF; Birk, JB; Frosig, C; Holten, M; Pilegaard, H; Dela F. 5'amp activated protein kinase expression in human skeletal muscle: Effects of strength training and type 2 diabetes. *J. Physiol.*, 2005, 564, 563–573.

[31] Birk, JB; Wojtaszewski, JF. Predominant α2/β2/γ3 ampk activation during exercise in human skeletal muscle. *J. Physiol.*, 2006, 577, 1021–1032.

[32] Frederich, M; Balschi, JA. The relationship between AMP-activated protein kinase activity and AMP concentration in the isolated perfused rat heart. *J Biol Chem.*, 2002 Jan 18, 277(3), 1928-32.

[33] Frederich, M; Zhang, L; Balschi, JA. Hypoxia and AMP independently regulate AMP- ctivated protein kinase activity in heart. *Am J Physiol Heart Circ Physiol*, 2005, 288, H2412–H2421.

[34] Zhou, G; Myers, R; Li, Y; Chen, Y; Shen, X; Fenyk-Melody, J; Wu, M; Ventre, J; Doebber, T; Fujii, N; Musi, N; Hirshman, MF; Goodyear, LJ; Moller, DE. Role of AMP-activated protein kinase in mechanism of metformin action. *J. Clin. Invest.*, 2001, 108, 1167–1174.

[35] Shaw, RJ; Lamia, KA; Vasquez, D; Koo, SH.; Bardeesy, N; Depinho, RA; Montminy, M; Cantley, LC. The kinase LKB1 mediates glucose homeostasis in liver and therapeutic effects of metformin. *Science*, 2005, 310, 1642–1646.

[36] Koo, SH; Flechner, L; Qi, L; Zhang, X; Screaton, RA; Jeffries, S; Hedrick, S; Xu, W; Boussouar, F; Brindle, P; Takemori, H;

Montminy, M. The CREB coactivator TORC2 is a key regulator of fasting glucose metabolism. *Nature*, 2005, 437, 1109–1111.

[37] He, L; Sabet, A; Djedjos, S; Miller, R; Sun, X; Hussain, MA; Radovick, S; Wondisford, FE. Metformin and insulin suppress hepatic gluconeogenesis through phosphorylation of CREB binding protein. *Cell*, 2009, 137, 635–646.

[38] Zhu, H; Moriasi, CM; Zhang, M; et al. Phosphorylation of serine 399 in LKB1 protein short form by protein kinase C is required for its nucleocytoplasmic transport and consequent AMP-activated proteinkinase (AMPK) activation. *J Biol Chem.*, 2013, 288, 16495-16505.

[39] Xie, Z; Dong, Y; Zhang, J; Scholz, R; Neumann, D; Zou, MH. Identification of the serine 307 of LKB1 as a novel phosphorylation site essential for its nucleocytoplasmic transport and endothelial cell angiogenesis. *Mol. Cell. Biol.*, 2009, 29, 3582–3596.

[40] Xie, Z; Dong, Y; Scholz, R; Neumann, D; Zou, MH. Phosphorylation of LKB1 at serine 428 by protein kinase C- is required for metformin-enhanced activation of the AMP-activated protein kinase in endothelial cells. *Circulation*, 2008, 117, 952–962.

[41] Dresser, MJ; Xiao, G; Leabman, MK; Gray, AT; Giacomini, KM. Interactions of n-tetraalkylammonium compounds and biguanides with a human renal organic cation transporter (hOCT2). *Pharm Res*, 2002, 19, 1244–1247.

[42] Wang, DS; Jonker, JW; Kato, Y; Kusuhara, H; Schinkel, AH; Sugiyama, Y. Involvement of Organic Cation Transporter 1 in Hepatic and Intestinal Distribution of Metformin. *J. Pharmacol. Exp. Ther.*, 2002, 302, 510–515.

[43] Kimura, N; Masuda, S; Tanihara, Y; Ueo, H; Okuda, M; Katsura, T; Inui, KI. Metformin Is a Superior Substrate for Renal Organic Cation Transporter OCT2 Rather Than Hepatic OCT1. *Drug Metab. Pharmacokinet.*, 2005, 20, 379–386.

[44] Han, T; Proctor, WR; Costales, CL; Cai, H; Everett, RS; Thakker, DR. Four Cation-Selective Transporters Contribute to Apical Uptake

and Accumulation of Metformin in Caco-2 Cell Monolayers. *J. Pharmacol. Exp. Ther.*, 2015, 352, 519–528.

[45] Graham, GG; Punt, J; Arora, M; Day, RO; Doogue, MP; Duong, JK; Furlong, TJ; Greenfield, JR; Greenup, LC; Kirkpatrick, CM; Ray, JE; Timmins, P; Williams, KM. Clinical Pharmacokinetics of Metformin. *Clin. Pharmacokinet.*, 2011, 50, 81–98.

[46] Masuda, S; Terada, T; Yonezawa, A; Tanihara, Y; Kishimoto, K; Katsura, T; Ogawa, O; Inui, KI. Identification and Functional Characterization of a New Human Kidney-Specific H+/Organic Cation Antiporter, Kidney-Specific Multidrug and Toxin Extrusion 2. *J. Am. Soc. Nephrol.*, 2006, 17, 2127–2135.

[47] Liang, X; Chien, HC; Yee, SW; et al. Metformin is a substrate and inhibitor of the human thiamine transporter, THTR-2 (SLC19A3). *Mol Pharm*, 2015, 12, 4301-4310.

[48] Urakami, Y; Okuda, M; Masuda, S; Saito, H; Inui, KI. Functional characteristics and membrane localization of rat multispecific organic cation transporters, OCT1 and OCT2, mediating tubular secretion of cationic drugs. *J Pharmacol Exp Ther*, 1998, 287, 800–805.

[49] Wu, X; Kekuda, R; Huang, W; Fei, YJ; Leibach, FH; Chen, J; Conway, SJ; Ganapathy, V. Identity of the organic cation transporter OCT3 as the extraneuronal monoamine transporter (uptake2) and evidence for the expression of the transporter in the brain. *J Biol Chem*, 1998, 273, 32776–32786.

[50] Sweet, DH; Pritchard, JB. OCT2 is a basolateral potential-driven carrier, not an organic cation/proton exchanger. *Am J Physiol*, 1999, 277, F890–F898.

[51] Zhou, M; Xia, L; Wang, J. Metformin Transport by a Newly Cloned Proton-Stimulated Organic Cation Transporter (Plasma Membrane Monoamine Transporter) Expressed in Human Intestine. *Drug Metab. Dispos.*, 2007, 35, 1956–1962.

[52] Lonsdale, D. A review of the biochemistry, metabolism and clinical benefits of thiamin(e) and its derivatives. Evid Based Complement *Alternat Med.*, 2006, 3(1), 49–59.

[53] Behal, RH; Buxton, DB; Robertson, JG; Olson, MS. Regulation of the pyruvate dehydrogenase multienzyme complex. *Annu Rev. Nutr.*, 1993, 13, 497-520.

[54] Chen, L; Shu, Y; Liang, X; Chen, EC; Yee, SW; Zur, AA; Li, S; Xu, L; Keshari, KR; Lin, MJ; Chien, HC; Zhang, Y; Morrissey, KM; Liu, J; Ostrem, J; Younger, NS; Kurhanewicz, J; Shokat, KM; Ashrafi, K; Giacomini, KM. OCT1 Is a High-Capacity Thiamine Transporter That Regulates Hepatic Steatosis and Is a Target of Metformin. *Proc. Natl. Acad. Sci. U. S. A.*, 2014, 111, 9983–9988.

[55] Donnino, M. Gastrointestinal beriberi: a previously unrecognized syndrome. *Ann. Intern Med.*, 2004, 141(11), 898–9. doi: 10.7326/0003-4819-141-11-200412070-00035.

[56] Duca, James; et al. Elevated Lactate Secondary to Gastrointestinal Beriberi. *Journal of general internal medicine*, 2016, 31, 1, 133-136. doi: 10.1007/s11606-015-3326-2.

[57] Romanski, SA; McMahon, MM. Metabolic Acidosis and Thiamine Deficiency. *Mayo Clin. Proc.*, 1999, 74, 259–263.

[58] Harding, JW; Jr. PE; Copeland, ES; White, HB. 3rd Role of glycerol 3-phosphate dehydrogenase in glyceride metabolism. Effect of diet on enzyme activities in chicken liver. *Biochem J.*, 1975, 146, 223–229.

[59] Harding, JW; Jr. PE; Morris, HP; White, HB. 3rd Proportional activities of glycerol kinase and glycerol 3-phosphate dehydrogenase in rat hepatomas. *Biochem J.*, 1975, 148, 545–550.

[60] Stryer, Lubert; Berg, Jeremy Mark; Tymoczko, John L. Biochemistry. San Francisco: W. H. Freeman 2007. ISBN 0-7167-8724-5.

[61] Madiraju, AK; Erion, DM; Rahimi, Y; Zhang, XM; Braddock, DT; Albright, RA; Prigaro, BJ; Wood, JL; Bhanot, S; MacDonald, MJ; Jurczak, MJ; Camporez, JP; Lee, HY; Cline, GW; Samuel, VT; Kibbey, RG; Shulman, GI. Metformin suppresses gluconeogenesis by inhibiting mitochondrial glycerophosphate dehydrogenase. *Nature.*, 2014, 510, 542–546.

[62] Hasenour, CM; et al. Mass spectrometry-based microassay of (2)H and (13)C plasma glucose labeling to quantify liver metabolic fluxes *in vivo*. *Am J Physiol Endocrinol Metab.*, 2015, 309, E191–203.

[63] Hunter, RW; Hughey, CC; Lantier, L; et al. Metformin reduces liver glucose production by inhibition of fructose-1-6-bisphosphatase. *Nat Med.*, 2018, 24(9), 1395–1406. doi: 10.1038/s41591-018-0159-7.

[64] Miller, RA; et al. Biguanides suppress hepatic glucagon signalling by decreasing production of cyclic AMP. *Nature*, 2013, 494, 256–260.

[65] Schommers, P; Thurau, A; Bultmann-Mellin, I; Guschlbauer, M; Klatt, AR; Rozman, J; Klingenspor, M; de Angelis, MH; Alber, J; Gründemann, D; Sterner-Kock, A; Wiesner, RJ. Metformin causes a futile intestinal-hepatic cycle which increases energy expenditure and slows down development of a type 2 diabetes-like state, *Molecular Metabolism*, 2017, doi: 10.1016/ j.molmet.2017.05.002.

[66] Wilcock, C; Bailey, CJ. Accumulation of metformin by tissues of the normal and diabetic mouse. *Xenobiotica.*, 1994, 24, 49–57. doi: 10.3109/00498259409043220.

[67] Bailey, CJ; Wilcock, C; Scarpello, JHB. Metformin and the intestine. *Diabetologia.*, 2008, 51, 1552–1553. doi: 10.1007/s00125-008-1053-5.

[68] Oh, JR; Song, HC; Chong, A; et al. Impact of medication discontinuation on increased intestinal FDG accumulation in diabetic patients treated with metformin. *AJR Am J Roentgenol.*, 2010, 195, 1404–1410. doi: 10.2214/AJR.10.4663.

[69] Capitanio, S; Marini, C; Sambuceti, G; Morbelli, S. Metformin and cancer: technical and clinical implications for FDG-PET imaging. *World J Radiol.*, 2015, 7, 57–60.

[70] Gontier, E; Fourme, E; Wartski, M; et al. High and typical 18F-FDG bowel uptake in patients treated with metformin. *Eur J Nucl Med Mol Imaging.*, 2008, 35, 95–99. doi: 10.1007/s00259-007-0563-6.

[71] Ozulker, T; Ozulker, F; Mert, M; Ozpacaci, T. Clearance of the high intestinal (18)F-FDG uptake associated with metformin after stopping the drug. *European Journal of Nuclear Medicine and Molecular Imaging.*, 2010, 37, 1011–1017.

[72] Abouzied, MM; Crawford, ES; Nabi, HA. 18F-FDG imaging: pitfalls and artifacts. *J Nucl. Med. Technol.*, 2005, 33, 145–155.

[73] Horakova, O; Kroupova, P; Bardova, K; Buresova, J; Janovska, P; Kopecky, J; Rosmeisl, M. *Metformin actuely lovers blood glucose levels by inhibition of intenstinal glucose transport Sci Rep.*, 2019, 9, 6156. Published online 2019 Apr 16. doi: 10.1038/s41598-019-42531-0.

[74] Bailey, CJ; Wilcock, C; Day, C. Effect of metformin on glucose metabolism in the splanchnic bed. *Br J Pharmacol.*, 1992, 105, 1009–1013.

[75] Davis, TM; Jackson, D; Davis, WA; Bruce, DG; Chubb, P. The relationship between metformin therapy and the fasting plasma lactate in type 2 diabetes: the Fremantle Diabetes Study. *Br J Clin Pharmacol.*, 2001, 52, 137–144.

[76] Qin, J; Li, Y; Cai, Z; et al. A metagenome-wide association study of gut microbiota in type 2 diabetes. *Nature.*, 2012, 490, 55–60. doi: 10.1038/nature11450.

[77] Tilg, H; Moschen, AR. Microbiota and diabetes: an evolving relationship. *Gut.*, 2014, 63, 1513–1521. doi: 10.1136/gutjnl-2014-306928.

[78] Forslund, K; Hildebrand, F; Nielsen, T; et al. Disentangling type 2 diabetes and metformin treatment signatures in the human gut microbiota. *Nature.*, 2015.

[79] Lee, H; Ko, G. Effect of metformin on metabolic improvement and gut microbiota. *Appl Environ Microbiol.*, 2014, 80, 5935–5943. doi: 10.1128/AEM.01357-14.

[80] Shin, NR; Lee, JC; Lee, HY; et al. An increase in the Akkermansia spp. population induced by metformin treatment improves glucose homeostasis in diet-induced obese mice. *Gut.*, 2014, 63, 727–735. doi: 10.1136/gutjnl-2012-303839.

[81] Scheen, AJ. Clinical pharmacokinetics of metformin. *Clin Pharmacokinet*, 1996, 30, 359–371.

[82] Bell, PM; Hadden, DR. Metformin. *Endocrinol Metab Clin North Am*, 1997, 199(26), 523–537.

[83] Lee, N; Duan, H; Hebert, MF; Liang, CJ; Rice, KM; Wang, J. Taste of a pill: organic cation transporter-3 (OCT3) mediates metformin accumulation and secretion in salivary glands. *J Biol Chem.*, 2014, 289, 27055–27064. doi: 10.1074/jbc.M114.570564.

[84] Proctor, WR; Bourdet, DL; Thakker, DR. Mechanisms underlying saturable intestinal absorption of metformin. *Drug Metab Dispos.*, 2008, 36, 1650–1658. doi: 10.1124/dmd.107.020180.

[85] Yee, SW; Lin, L; Merski, M; et al. Prediction and validation of enzyme and transporter off-targets for metformin. *J Pharmacokinet Pharmacodyn.*, 2015, 42, 463–475.

[86] Dandona, P; Fonseca, V; Mier, A; Beckett, AG. Diarrhea and metformin in a diabetic clinic. *Diabetes Care.*, 1983, 6(5), 472–474.

[87] Bouchoucha, M; Uzzan, B; Cohen, R. Metformin and digestive disorders. *Diabetes Metab.*, 2011, 37(2), 90–96. doi, 10.1016/j.diabet.2010.11.002.

[88] Hundal, RS; Krssak, M; Dufour, S; et al. Mechanism by which metformin reduces glucose production in type 2 diabetes. *Diabetes.*, 2000, 49(12), 2063–2069. doi: 10.2337/diabetes.49.12.2063.

[89] Stocker, SL; Morrissey, KM; Yee, SW; Castro, RA; Xu, L; Dahlin, A; Ramirez, AH; Roden, DM; Wilke, RA; McCarty, CA; Davis, RL; Brett, CM; Giacomini, KM. (2013) The effect of novel promoter variants in MATE1 and MATE2 on the pharmacokinetics and pharmacodynamics of metformin. *Clin. Pharmacol. Ther.*, 93, 186–194.

[90] Tucker, GT; Casey, C; Phillips, PJ; Connor, H; Ward, JD; Woods, HF. (1981). Metformin kinetics in healthy subjects and in patients with diabetes mellitus. *Br. J. Clin. Pharmacol.*, 12, 235–246.

[91] Magnusson, I; Rothman, DL; Jucker, B; Cline, GW; Shulman, RG; Shulman, GI. Liver glycogen turnover in fed and fasted humans. *Am J Physiol*, 1994, 266 (Endocrinol Metab 29), E796–E803.

[92] Roden, M; Perseghin, G; Petersen, KF; Hwang, JH; Cline, GW; Gerow, K; Rothman, DL; Shulman, GI. The roles of insulin and glucagon in the regulation of hepatic glycogen synthesis and turnover in humans. *J Clin Invest*, 1996, 97, 642–648.

[93] Petersen, KF; Laurent, D; Rothman, DL; Cline, GW; Shulman, GI. Mechanism by which glucose and insulin inhibit net hepatic glycogenolysis in humans. *J Clin Invest*, 1998, 101, 1203–1209.

[94] DeFronzo, RA; Barzilai, N; Simonson, DC. Mechanism of metformin action in obese and lean non-insulin-dependent diabetic subjects. *J Clin Endocrino Metab*, 1991, 73, 1294–1301.

[95] Wu, MS; Johnson, P; Shen, W-H; Hollenbeck, C; Jeng, C; Goldine, I; Chen, YD; Reaven, G. Effect of metformin on carbohydrate and lipoprotein metabolism in NIDDM patients. *Diabetes Care*, 1990, 13, 1–8.

[96] Perriello, G; Misericordia, P; Volpi, E; Santucci, C; Ferrannini, E; Ventura, M; Santeusiano, F; Brunetti, P; Bolli, G. Acute antihyperglycemic mechanisms of metformin in NIDDM: evidence for suppression of lipid oxidation and hepatic glucose production. *Diabetes*, 1994, 43, 920–928.

[97] Radziuk, J; Zhang, Z; Wiernsperger, N; Pye, S. Effects of metformin on lactate uptake and gluconeogenesis in the perfused rat liver. *Diabetes*., 1997, 46, 1406-13.

[98] Stumvoll, M; Nurjhan, N; Perriello, G; Dailey, G; Gerich, JE. Metabolic effects of metformin in non-insulin-dependent diabetes mellitus. *N Engl J Med*., 1995, 333, 550–554.

[99] Schafer, G. Biguanides. A review of history, pharmacodynamics and therapy. *Diabetes Metab*., 1983, 9, 148–163.

[100] Pryor, HJ; Smyth, JE; Quinlan, PT; Halestrap, AP. Evidence that the flux control coefficient of the respiratory chain is high during gluconeogenesis from lactate in hepatocytes from starved rats. *Biochem. J*., 1987, 247, 449–457.

[101] Cao, J; Meng, S; Chang, E; et al. Low concentrations of metformin suppress glucose production in hepatocytes through AMP-activated protein kinase (AMPK). *J Biol Chem*., 2014, 289(30), 20435–20446. doi: 10.1074/jbc.M114.567271.

[102] Koopmans, SJ; Ohman, L; Haywood, JR; Mandarino, LJ; DeFronzo, RA. Seven days of euglycemic hyperinsulinemia induces insulin resistance for glucose metabolism but not hypertension, elevated

catecholamine levels, or increased sodium retention in conscious normal rats. *Diabetes.*, 1997, 46, 1572–1578. doi: 10.2337/ diacare.46.10.1572.

[103] Van Loon, LJC; Koopman, R; Manders, R; Van der Weegen, W; Gerrit, P; Van Kranenburg, GP; Keizer, HA. Intramyocellular lipid content in type 2 diabetes patients compared with overweight sedentary men and highly trained endurance athletes *Am J Physiol Endocrinol Metab*, 2004287, 558–565.

[104] Bailey, CJ; Puah, JA. Effect of metformin on glucose metabolism in mouse soleus muscle. *Diabetes Metab*, 1986, 12, 212–218.

[105] Liu, Y; Wan, Q; Guan, Q; Gao, L; Zhao, J. High-fat diet feeding impairs both the expression and activity of AMPKa in rats' skeletal muscle. *Biochem Biophys Res Commun*, 2006, 339, 701–707.

[106] Hundal, HS; Ramlal, T; Reyes, R; Leiter, LA; Klip, A. Cellular mechanism of metformin action involves glucose transporter translocation from an intracellular pool to the plasma membrane in L6 muscle cells. *Endocrinology.*, 1992, 131, 1165–1173.

[107] Musi, N; Hirshman, MF; Nygren, J; Svanfeldt, M; Bavenholm, P; Rooyackers, O; Zhou, G; Williamson, JM; Ljunqvist, O; Efendic, S; et al. Metformin increases amp-activated protein kinase activity in skeletal muscle of subjects with type 2 diabetes. *Diabetes.*, 2002, 51, 2074–2081.

[108] Russell, RR; 3rd, Bergeron, R; Shulman, GI; Young, LH. Translocation of myocardial GLUT-4 and increased glucose uptake through activation of AMPK by AICAR. *Am J Physiol Heart Circ Physiol*, 1999, 277, H643–H649.

[109] Friedman, JM; Halaas, JL. Leptin and the regulation of body weight in mammals. *Nature.*, 1998, 395, 763–770. doi: 10.1038/27376.

[110] Tokubuchi, I; Tajiri, Y; Iwata, S; et al. Beneficial effects of metformin on energy metabolism and visceral fat volume through a possible mechanism of fatty acid oxidation in human subjects and rats. *PLoS One.*, 2017, 12(2), e0171293. Published 2017 Feb 3. doi: 10.1371/journal.pone.0171293.

[111] Aghili, R; Malek, M; Valojerdi, AE; Banazadeh, Z; Najafi, L; Khamseh, ME. Body composition in adults with newly diagnosed type 2 diabetes: effects of metformin. *J Diabetes Metab Disord.*, 2014, 13(1), 88 Epub 2014/09/24. PubMed Central PMCID: PMC4159548. 10.1186/s40200-014-0088-z.

[112] Hu, Y; Young, AJ; Ehli, EA; Nowotny, D; Davies, PS; Droke, EA; et al. Metformin and berberine prevent olanzapine-induced weight gain in rats. *PLoS One.*, 2014, 9(3), e93310 PubMed Central PMCID: PMCPMC3965561. 10.1371/journal.pone.0093310 [PMC free article] [PubMed] [Cross Ref] [Google Scholar].

[113] Jansson, PA1; Gudbjörnsdóttir, HS; Andersson, OK; Lönnroth, PN. *The Effect of Metformin on Adipose Tissue Metabolism and Peripheral Blood Flow in Subjects With NIDDM. Diabetes Care*, Feb 1996, 19 (2), 160-164. DOI: 10.2337/diacare.19.2.16.

[114] Kim, HJ; Park, EY; Oh, MJ; Park, SS; Shin, KH; Choi, SH; et al. Central administration of metformin into the third ventricle of C57BL/6 mice decreases meal size and number and activates hypothalamic S6 kinase. *Am J Physiol Regul Integr Comp Physiol.*, 2013, 305(5), R499–505. Epub 2013/07/05. 10.1152/ajpregu. 00099.2013.

[115] Lee, CK; Choi, YJ; Park, SY; Kim, JY; Won, KC; Kim, YW. Intracerebroventricular injection of metformin induces anorexia in rats. *Diabetes Metab J.*, 2012, 36(4), 293–9. Epub 2012/09/06. PubMed Central PMCID: PMC3428418. 10.4093/dmj.2012. 36.4.293.

[116] Lee, AJ. Metformin in noninsulin-dependent diabetes mellitus. *Pharmacotherapy.*, 1996, 16(3), 327–51. Epub 1996/05/01.

[117] Helvaci, MR; Kaya, H; Borazan, A; Ozer, C; Seyhanli, M; Yalcin, A. Metformin and parameters of physical health. *Intern Med.*, 2008, 47(8), 697–703. Epub 2008/04/19.

[118] Sun, EW; Martin, AM; Wattchow, DA; de Fontgalland, D; Rabbitt, P; Hollington, P; Young, RL; Keating, DJ. Metformin Triggers PYY Secretion in Human Gut Mucosa. *J Clin Endocrinol Metab.*, 2019 Jul 1, 104(7), 2668-2674. doi: 10.1210/jc.2018-02460.

[119] Duca, FA; Côté, CD; Rasmussen, BA; et al. Metformin activates a duodenal AMPK-dependent pathway to lower hepatic glucose production in rats. *Nat Med.*, 2015, 21, 506–511.

[120] Salpeter, SR; Greyber, E; Pasternak, GA; Salpeter, EE (November 2003). "Risk of fatal and nonfatal lactic acidosis with metformin use in type 2 diabetes mellitus: systematic review and meta-analysis". *Archives of Internal Medicine.*, 163 (21), 2594–602.

[121] Nathan, DM; Buse, JB; Davidson, MB; Ferrannini, E; Holman, RR; Sherwin, R; Zinman, B. Medical management of hyperglycaemia in type 2 diabetes mellitus: a consensus algorithm for the initiation and adjustment of therapy: a consensus statement from the American Diabetes Association and the European Association for the Study of Diabetes. *Diabetologia.*, 2009, 52(1), 17–30.

[122] Stang, M; Wysowski, DK; Butler-Jones, D. Incidence of lactic acidosis in metformin users". *Diabetes Care.*, 1999, 22(6), 925–927.

[123] Shepherd, P; Kahn, B. Glucose transporters and insulin action. *N Engl. J. Med.*, 1999, 341, 240-246.

[124] Donohue, WL. Dysendocrinism. *J. Pediat.*, 1948, 32, 739-748.

[125] Donohue, WL; Uchida, IA. Leprechaunism: a euphemism for a rare familial disorder. *J Pediat.*, 1954, 45, 505-519.

[126] Rabson, S; Mendenhall, E. Familial hypertrophy of pineal body, hyperplasia of adrenal cortex and diabetes mellitus; report of 3 cases. *Am J Clin Pathol.*, 1956, 26(3), 283–290. PMID 13302174.

[127] De Fronzo, RA. Glucose clamp technique: a method for quantifying insulin secretion and resistence. *Am J Physiol*, 1979, 237, 214-223.

[128] Reaven, GM. Role of insulin resistance in human disease. *Diabetes*, 1988, 37, 1595-1607.

[129] Kubat, K. Paradox of insulin resistance. *Vnitr. Lek.*, 1999, 10, 614-617.

[130] Friedman, JE; Dohm, GL; Leggett-Frazier, N; Elton, CW; Tapscot, EB; Pories, WP; Caro, J. Restoration of insulin responsiveness in skeletal muscle of morbidly obese patients after weight loss. *J Clin Invest*, 1992, 89, 701-705.

[131] Pories, WJ; MacDonald, KG; Jr. Morgan, EJ; et al. Surgical treatment of obesity and its effect on diabetes. 10-y follow up. *Am J Clin Nutr*, 1992, 55(Suppl.), 582S-585S.

[132] Polyzogopoulou, EV; Kalfarentzos, F; Vagenakis, AG; Alexandrides, TK. Restoration of euglycemia and normal acute insulin response to glucose in obese subject with type 2 diabetes following bariatric surgery. *Diabetes*, 2003, 52, 1098-1103.

[133] Schilling, EE; Rechler, MM; Grunfeld, C; Rosenberg, AM. Primary defect of insulin receptors in skin fibroblasts cultured from an infant with leprechaunism and insulin-resistance. *Proc Nat Acad Sci.*, 1979, 76, 5877-81.

[134] Reddy, SSK; Lauris, V; Kahn, CR. Insulin receptor function in fibroblasts from patients with leprechaunism: differential alterations in binding, autophosphorylation, kinase activity, and receptor-mediated internalization. *J. Clin. Invest.*, 1988, 82, 1359-65.

[135] Kubat, K. Model of Diabetes Mellitus Type 2, T2DM. *J Nutr Food Sci*, 5, 2015, 344. doi: 10.4172/2155-9600.1000344.

[136] Groop, LC; Tuomi, T. Non-insulin-dependent diabetes mellitus - a collision between thrifty genes and an affluent society. The Finnish medical society duodecim, *Ann med*, 1997, 29, 37-53.

[137] Neel, JV. Diabetes mellitus: A "thrifty" genotype rendered detrimental by "progress"? *Am J Hum Genet*, 1962, 14, 353-352

[138] Van Loon, LJC; Koopman, R; Manders, R; Van der Weegen, W; Gerrit, P; et al. Intramyocellular lipid content in type 2 diabetes patients compared with overweight sedentary men and highly trained endurance athletes. *Am J Physiol Endocrinol Metab*, 2004, 287, 558-565.

[139] Mokdad, AH; Ford, ES; Bowman, BA; et al. Prevalence of obesity, diabetes and obesity-related health risk factors, 2001. *JAMA*, 2003, 289, 76-79.

[140] Diabetes Prevention Program Research Group. Reduction in the Incidence of Type 2 Diabetes with Lifestyle Intervention or Metformin. *The New England journal of medicine*, 2002, 346(6), 393-403.

[141] Effect of intensive blood-glucose control with metformin on complications in overweight patients with type 2 diabetes (UKPDS 34). UK Prospective Diabetes Study (UKPDS) Group. *Lancet*, 1998 Sep 12, 352(9131), 854-865.

[142] Kooy, A; de Jager, J; Lehert, P; Bets, D; Wulffele, MG; Donker, AJ; Stehouwer, CD. Long-term effects of metformin on metabolism and microvascular and macrovascular disease in patients with type 2 diabetes mellitus. *Arch. Intern. Med.*, 2009, 169, 616–625.

[143] Younossi, ZM; et al. Global epidemiology of nonalcoholic fatty liver disease-Meta-analytic assessment of prevalence, incidence, and outcomes. *Hepatology*, 2016, 64, 73–84, https://doi.org/10.1002/hep.28431.

[144] Li, Y; Liu, L; Wang, B; Wang, J; Chen, D. Metformin in non-alcoholic fatty liver disease: A systematic review and meta-analysis. *Biomedical Reports*, 2013, 1, 57-64. https://doi.org/ 10.3892/br.2012.18.

[145] Seifarth, C1; Schehler, B; Schneider, HJ. Effectiveness of metformin on weight loss in non-diabetic individuals with obesity. *Exp Clin Endocrinol Diabetes.*, 2013 Jan, 121(1), 27-31.

[146] Park, MH; Kinra, S; Ward, KJ; et al. Metformin for obesity in children and adolescents: a systematic review. *Diabetes Care*, 2009, 32, 1743-1745.

[147] Mauras, N; DelGiorno, C; Hossain, J; et al. Metformin use in children with obesity and normal glucose tolerance-effects on cardiovascular markers and intrahepatic fat. *J Pediatr Endocrinol Metab*, 2012, 25, 33-40.

[148] Pasquali, R; Gambineri, A; Biscotti, D; Vicennati, V; Gagliardi, L; Colitta, D; et al. Effect of long-term treatment with metformin added to hypocaloric diet on body composition, fat distribution, and androgen and insulin levels in abdominally obese women with and without the polycystic ovary syndrome. *J Clin Endocrinol Metab.*, 2000, 85, 2767-74.

[149] Yi, Y; Zhang, W; Yi, J; Xiao, ZX. Role of p53 Family Proteins in Metformin Anti-Cancer Activities. *J Cancer.*, 2019, 10(11), 2434–2442. doi:10.7150/jca.30659.

[150] Bailey, CJ. Metformin: historical overview. *Diabetologia*, 2017, 60, 1566 1576.

In: The Pharmacological Guide to Metformin ISBN: 978-1-53616-634-7
Editor: Natalie Lajoie

Chapter 4

METFORMIN EFFECTS IN CANCER MODELS

André Gustavo de Oliveira[1], PhD
and Maria Cristina C. Gomes-Marcondes[2,*], PhD
[1]Obesity and Comorbidities Research Centre, Department of Structural and Functional Biology, Institute of Biology, University of Campinas, UNICAMP, Campinas, SP, Brazil
[2]Laboratory of Cancer and Nutrition, Department of Structural and Functional Biology, Institute of Biology, University of Campinas, UNICAMP, Campinas, SP, Brazil

ABSTRACT

Metformin remains the first choice and the most prescribed medicine for the treatment of diabetes. This pharmacological drug is effective in reducing glucose levels, has a low cost and causes mild side effects in a few numbers of patients. The reason why diabetic patients are prone to develop cancer is the combination of insulin resistance, high insulin circulating levels and increased insulin-like growth factors, all of them are well known in promoting and sustaining the growth of various type of tumours. In fact, metformin has other effects than decreasing both the

* Corresponding Author's E-mail: cintgoma@unicamp.br.

insulin and IGF levels. It also acts in killing the tumour cells and consequently decreasing the tumour mass. A large number of studies both *in vitro* and *in vivo* have demonstrated that metformin treatment can lead tumour cells to anti-proliferative, autophagy and/or apoptosis processes. Since cancer induces cachexia state, which leads to an involuntary body mass waste, mainly muscle protein, we have seen previously that metformin could also recover some changed parameters on protein and glucose metabolism in an experimental model of cachexia, improving the host responses in these tumour-bearing rats. In this present chapter, we are going to explore the differences between healthy and cancer cells and how this can influence the metformin effect on these cells.

Metformin and Cancer – A Brief History of Their Association

Cancer is the second pathology with the highest mortality rate in the world, behind only cardiovascular diseases, and the World Health Organisation shows an increased number of new cases and also deaths every year ("World Health Organization - International Agency for Research on Cancer" 2014). Cancer promotes significant changes in host metabolism, both by compromising the function of the organ in which it develops and by changing the metabolic and biosynthetic pathways as well. The tumour development demands high energetic expenses, associated with higher inflammatory response in the host, which could lead to a cachexia state, responsible for the spoliation of fat reserve tissues, muscle tissue and also for the hypertrophy of organs compromised by the generalised and chronic host inflammatory response (Argilés et al. 2007; W. J. Evans et al. 2008). According to some studies definition, cancer patients could be in a cachectic state having at least 5% body weight loss added to at least three of the following symptoms (i) asthenia, (ii) fatigue, (iii) anorexia, (iv) low lean mass index, (v) biochemical changes (such as increased C-reactive protein concentration, IL-6, anemia (hemoglobin less than 12 g/dL) or (vi) decreased serum albumin concentration (less than 3.2 g/dL), and this situation is the most frequent cause of morbidity and mortality (W. J. Evans et al. 2008; Tisdale 2002; Choudry et al. 2006).

The association between type 2 diabetes mellitus (T2DM) and increased cancer incidence has been described since 1934 (MARBLE 1934). Diabetes is considered an important predictor and a prognostic factor for different types of cancer such as colon, pancreatic, breast, liver and bladder cancer (Coughlin et al. 2004). Diabetic patients show high levels of circulating insulin due to decreased insulin sensibility in peripheral tissue such as muscle and with adipose tissue; therefore, the increased insulin secretion is a compensatory effect to the decreased insulin sensibility. Therefore, the increased insulin level might be a link between diabetes and cancer (Larsson, Mantzoros, and Wolk 2007; Vigneri et al. 2009; Pollak 2008).

Although metformin is the most used drug to treat diabetic patients, it has only been prescribed since the 1950s (Bailey 2017). Only in the last decade, the association between metformin and reduced cancer incidence in diabetic patients had been proposed (J. M. M. Evans et al. 2005). Studies have shown that the incidence of cancer in patients taking metformin was decreased (Libby et al. 2009; Decensi et al. 2010; J. M. M. Evans et al. 2005) or, in the occurrence of cancer, had a better prognosis (Libby et al. 2009). These associations seem reasonable since metformin acts mainly (i) decreasing the hepatic glucose production (namely gluconeogenesis), (ii) increasing the sensibility of insulin receptor to insulin (acting as "sensitiser agent"), (iii) decreasing the circulating insulin level. How these metformin's effects impact on tumour cells will be the subject of this chapter and whenever appropriate the differences with healthy tissues will be discussed. Another feature of metformin is the activation of AMP-dependent protein kinase (AMPK) in muscle, fat, and liver cells (Zhou et al. 2001). The increased AMPK activity promoted by metformin results in stimulation of glucose uptake and fatty acid oxidation, inhibition of hepatic glucose production, cholesterol, triacylglycerol synthesis and lipogenesis (Ruderman et al. 2003), which decreases the AMP/ATP ratio, and thus performing the cellular homeostasis in the peripheral tissues of diabetic patients.

Metformin activates AMPK both *in vivo* and *in vitro* (Zhou et al. 2001) and many experiments have described its therapeutic potential for mTOR inhibition, inducing anti-proliferative and anti-angiogenic effects in such as treating cancer patients, heart disease, autoimmune and metabolic (Wullschleger, Loewith, and Hall 2006; Inoki, Corradetti, and Guan 2005; Manning et al. 2002). The antiproliferative effect of metformin was demonstrated using MCF-7, BT-474 and SKBR-3 cells (human breast cancer-derived cell lines) and also in experimental model of cachexia, the Walker-256 cells, in which metformin inhibited cell proliferation, colony formation and caused, at least partially, the cell cycle arrested to the G0 phase (Alimova et al. 2009).

Additionally, metformin decreases circulating insulin levels, inhibits angiogenesis and exerts a toxic effect on tumour cells (Kourelis and Siegel 2012). Our findings also corroborate these facts, as Walker-256 cultured cells also slowed the growth in the metformin-treated group. We also demonstrated that metformin could reduce the relative weight of the tumour mass, consistent with the smaller number of positively Ki-67 labelled cells, and also reduced PCNA expression in tumour-bearing rats treated with metformin. On the other hand, *in vitro* experiments, we observed high doses of metformin (100 mM) reduced the tumour cells viability in just 24 hours, and lower doses of metformin (1 mM) also reduced cell viability in a longer time (72 hours). Besides, the proliferative capacity of Walker-256 cells was significantly reduced by metformin treatment, decreasing the DNA synthesis rate and the ability to form colonies (data not yet published).

The reason for the decrease in both metformin-modulated neoplastic cell proliferation and viability was possibly due to AMPK phosphorylation by a mechanism not yet elucidated. AMPK is considered the intracellular energy status sensor, i.e., this protein is capable of detecting variations in ATP concentration and regulates positively or negatively the energy processes (which consume energy) or degradation (which produce energy). Thus, probably the energy stress imposed by the tumour development, evidenced by the lower plasma glucose concentration, could be due to the metformin action promoting an increase in AMPK phosphorylation, which inhibited the activation of mTOR, the key protein for all cell proliferation process. As previously demonstrated, metformin treatment also modulated neoplastic growth by reducing activation of the IRS-1/PI3K/Akt/mTOR pathway in tumour cells, both *in vivo* and *in vitro*. At the same time, different types of stress induce phosphorylation and consequent stabilisation of p53 (Maclaine and Hupp 2009). The p53 protein is known as the "genome guardian" because it prevents replicating of damaged DNA causing diseases like cancer. P53 is usually associated with Mdm2 and tagged to degradation through proteasome, maintaining its low concentration within the cell and allowing the cell to proliferate. When DNA damage occurs, there is p53 phosphorylation at different amino acid residues, which no longer has an affinity for Mdm2, leading to its accumulation within the cell and then acting as a transcription factor. Its stabilisation initiates various cellular processes in order to eliminate damage and allow the cell to survive. These include damaged DNA repair, cell cycle suspension and senescence (Menendez, Inga, and Resnick 2009). If the DNA damage is irreparable, p53 is also responsible for initiating the process of apoptotic cell death. Indeed, in our previous results, we found that metformin treatment increased p53 phosphorylation and also the expression of the pro-apoptotic Bax protein while decreased the expression of the anti-apoptotic protein Bcl-2 in Walker 256 tumour.

Indeed, the effect of metformin occurred primarily by the maintenance of serum insulin and IGF-1 concentration, concurrently with increased AMPK phosphorylation and possibly by increased TSC2 phosphorylation, which is inhibited by Akt. In this way, metformin treatment modulated the neoplastic growth by reducing activation of the IRS-1/PI3K/Akt/mTOR pathway in the tumour cells. As a result, there is less mTOR phosphorylation, and consequently, there is lower protein synthesis in the tumour cells. This result was confirmed by incubating tumour tissue with metformin, in which we observed less incorporation of H^3-phenylalanine, associated with increased AMPK phosphorylation state. Thus, from these findings, both *in vivo* and *in vitro*, metformin treatment promoted inhibition of protein synthesis consistent with lower cell proliferation and lower tumour evolution (lower Ki-67 and PCNA labelling ratio) in this Walker-256 tumour, which is an experimental model of cachexia.

Solid tumours usually have a high demand for glucose due to the characteristic described by Otto Warburg over 80 years ago (WARBURG 1956). Named as the Warburg effect, this happens because solid tumours have poorly vascularized tissue portions and, consequently, the oxygen diffusion rate does not provide a rapid supply to allow maintenance of glucose metabolism via oxidative phosphorylation(Hammoudi et al. 2011). To minimise this problem, tumour cells oxidise high glucose rates, preferably by glycolysis to lactate, with low ATP production, providing a high energy demand to the host (Denko 2008). Our previous results corroborate this fact since Walker tumour growth promoted great hypoglycaemia, associated with higher lactate serum concentration, this process leads to a stimulus for increase the gluconeogenesis rate, in some way to maintain a normoglycaemia in the host. Consequently, there is a deeply negative energy cycle, since the gluconeogenesis consumes three ATP molecules for each glucose molecule produced, including the release of gluconeogenic precursors in the Cori cycle such as alanine originated from muscle protein degradation. Thus, the increase in gluconeogenesis rate in cancer patients represents an additional expenditure of 300 Kcal/day (Edén et al. 1984).

The skeletal muscle is deeply wasted in a tumour-bearing host, not only by the futile Cori cycle but also by the lower serum insulin concentration. This fact, associated with the higher glucagon concentration could partially explain the lower protein synthesis rate, analysed by reduced activation of Akt and mTOR in the gastrocnemius muscle in the Walker-256 tumour-bearing animals (A. G. Oliveira and Gomes-Marcondes 2014; André G. Oliveira and Gomes-Marcondes 2016). The lower insulin concentration caused a decrease in glucose uptake by the muscle, which consequently entered into a state of energy stress, further aggravated by the low availability of this substrate (reduction in serum glucose). Consequently, phosphorylated AMPK inhibited the protein synthesis by inhibition of mTOR and also decreasing the FoxO3a phosphorylation, which induced higher transcription of genes responsible for the synthesis of two important enzymes related to protein degradation, like MuRF-1 and atrogin (Gumucio and Mendias 2013). The consequence of this activation of AMPK, leading the protein degradation, was likely to re-establish the muscle ATP levels.

The process of muscle protein degradation occurs mainly by the action of lysosomal, calcium-dependent and proteasomal pathway enzymes (Hasselgren, Wray, and Mammen 2002), and our previous data showed that the Walker tumour increased the enzyme activity of all these pathways except the calcium-dependent pathway (André G. Oliveira and Gomes-Marcondes 2016). Chymotrypsin-like enzyme activity reflects the proteasomal degradation pathway, and it is mainly affected by cancer-cachexia-induced muscle waste. The mechanism by which the tumour increases the expression and activity of the proteasomal subunits is not fully known, but it is known that tumour products can modulate this process (Tisdale 2009; Mendes et al. 2015). Proteolysis-inducing factor (PIF) is described as a protein secreted by cachexia-inducing tumours that is capable of promoting proteolysis in host muscle cells. Although it has not yet been proved that Walker 256 tumours produce the same protein, studies from our group have shown that Walker 256 tumours also produce a protein very similar to PIF (Yano et al. 2008). This Walker 256 tumour-derived protein, called PIF-like or Walker Factor, may have been responsible for increasing protein degradation in the gastrocnemius muscle, as observed by the decreased relative weight of this tissue, as well as decreased synthesis, analysed by phosphorylation of mTOR (Cruz et al. 2019).

THE MITOGENIC BRANCH OF THE INSULIN PATHWAY

Besides the insulin effects over glucose, lipid and protein metabolism, insulin is also well established as a mitogen molecule, being related to cell growth, division, migration and apoptosis (Sasaoka et al. 1994). Many studies have demonstrated that insulin and insulin-analogues have potential to work as a tumour growth factor (Gupta et al. 2002; Hemkens et al. 2009; Jonasson et al. 2009; Draznin 2010).

The main pathway involved in insulin signalling is the RAS/RAF/MEK/ERK axis. Erk (also known as MAPK) is a protein kinase that mediates the binding of growth factors to cell membrane receptors and signalling transduction. The MEK (also known as MAPKK) and Erk signalling cascade are related to cell transformation and tumour progression (Shaul and Seger 2007) (Hoshino et al. 1999). The ErK signalling pathway is abnormally stimulated in various types of tumours as it is responsible for the phosphorylation of myosin light chain kinase, calpains, focal adhesion kinase and paxillin (Huang et al. 2004) and these proteins promote the migration of neoplastic cells (Kim and Choi 2010). On the other hand, the Erk signalling pathway also regulates the activity and expression of Bcl-2 family proteins, leading to neoplastic cell survival (Balmanno and Cook 2009). Growth factors such as IGF-1 and EGF (epidermal growth factor) stimulate receptor tyrosine kinase activity (RTK-Grb2-SOS), which phosphorylate and activate the Ras, and consequently activate Raf. The Raf proteins (A-Raf, B-Raf or C-Raf) are responsible for phosphorylating and activating MEK, and this kinase directly phosphorylates Erk1 or Erk2 (Erk1/2) (Dhillon et al. 2007; Malumbres and Barbacid 2003).

In this sense, the increased in the insulin sensibility by some tissues caused by metformin is also a good explanation of how metformin can decrease some tumours growth rate. However, in our previous studies, we have shown that metformin treatment modulated neoplastic growth by reducing activation of the IRS-1/PI3K/Akt/mTOR pathway, the main effect of metformin occurred maintaining serum insulin and IGF-1 content, which in parallel could also improve the host tissues, especially skeletal muscle mass (André G Oliveira and Gomes-Marcondes 2016). At the same time, the hormonal body milieu is responsible for the growth processes as well as the balance of this process, also keeping the host tissues.

Insulin and IGF-1 are mitogenic hormones and growth factors, but also anabolic hormones that maintain the healthier features of all host tissues and, as such, induce cells to grow in mass and also in number. The IRS-1/PI3K/Akt/mTOR signalling pathway is responsible for this process and is controlled by these hormones. The binding of IGF-1 to its receptor on the cell membrane leads to autophosphorylation of tyrosine residues and consequently to insulin receptor substrate (IRS-1) phosphorylation. In our previous study in skeletal muscle, we observed a reduction on PI3K phosphorylation and activation and, subsequently, Akt phosphorylation and activation in muscle cells of Walker-256 tumour-bearing rats. Although our findings showed that metformin-treated tumour-bearing group reduced the total IRS-1 concentration, this effect was associated with the similar IGF-1 concentration, since these metformin-treated tumour-bearing animals had an improvement of protein balance in the gastrocnemius muscle, which likely corresponded to signalling pathways for protein synthesis that may be regulated by other mechanisms. This effect is possibly due to the lack of IGF-1 action, since the activation of AMPK promoted by metformin indirectly inhibited the insulin-stimulated cell proliferation, in the case of neoplastic cells (Vigneri et al. 2009) (McCarty 2004), which consequently reduced PIF-like production and secretion. This fact could explain the maintenance of lean body mass in metformin-treated tumour-bearing animals, as well as the results of the muscle signalling pathways (Figure 1). Thus, with the decrease in tumour growth, probably there was less release of PIF-like, which could minimise the activation of proteolytic pathways (the 20S, 19S and 11S subunits) and thus preserving muscle mass. This fact was even more beneficial since treatment with metformin also provided the maintenance of serum insulin concentration and IGF-1, factors directly responsible for the activation of cellular signalling for protein synthesis in the muscle. On the other hand, parallel to the increase in synthesis, there was a decrease in the degradation process, observed by the reduction of FoxO translocation and consequently reducing ubiquitin ligases, thus minimising muscle spoliation in metformin-treated animals.

Mitochondria and Its Importance in Cell Fate

Although metformin is the most prescribed treatment to patients with type 2 diabetes (T2DM), its mechanism of action is still not completely understood. The most accepted idea is that metformin act mainly by inhibiting the mitochondrial respiratory complex I (Owen, Doran, and Halestrap 2000; Wheaton et al. 2014; Ling et al. 2017). Metformin is a positively charged molecule (cation) and is actively transported into the mitochondrial organelle, tending to accumulate in the matrix where it is in the locality of complex I. The inhibition in the complex I promoted by metformin causes a decrease in the ATP production, leading to an increase in the AMP:ATP ratio. The raised AMP level inside the cell is a potent indicator of energy stress. The AMP:ATP ratio is tighten controlled by the activity of the adenylate kinase, which reaction can be written 2ADP ↔ ATP + AMP, and is maintained close to equilibrium. Cells respond in a greater extend to changes in the AMP concentration than to ADP or ATP, making the AMP a much better sensor of the energy status inside the cells (Hardie and Hawley 2001). The increased AMP concentration leads to activation of the AMP-activated protein kinase (AMPK), which is considered one of the master regulators of metabolic pathways (Garcia and Shaw 2017). In summary, the effect of metformin is an indirect activation of AMPK (Figure 1), since there is no evidence supporting the direct action in activating AMPK.

The inhibition of electron transport chain is a source of reactive oxygen species (ROS) (Chen et al. 2007), and an imbalance between ROS production and degradation leads cells to a state known as oxidative stress. In such conditions, the cell can increase the autophagic flux trying to restore the usual electron flux over the electron transport chain, or if the damage is too much to be repaired, cells can enter in a death process, such as necrosis or apoptosis.

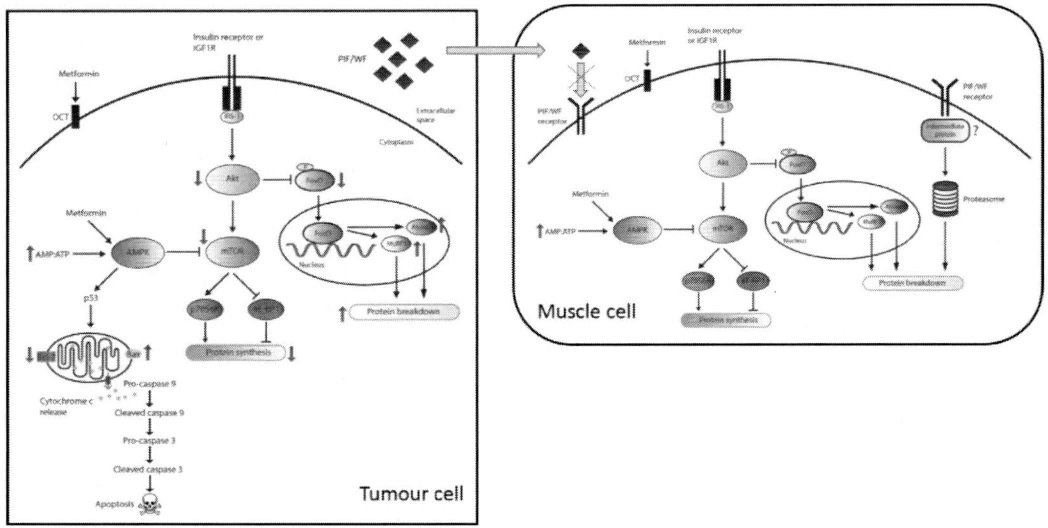

Figure 1. Scheme showing the metformin action on inducing the tumour cells death and also maintaining the skeletal muscle mass in cancer-cachexia model. Metformin treatment increases the tumour cell intracellular concentration of AMP (3',5' adenosine monophosphate), which leads to AMPK (AMP-activated protein kinase) activation. This effect causes an inhibition of mTOR (mechanistic target of rapamycin) pathway and consequently, inhibition of protein synthesis. In addition, the decreased circulating levels of insulin and IGF-1 (insulin-like growth factor 1) results in a decreased activation of Akt (protein kinase B) and decreased mTOR activation. The reduction of Akt activation also decreases the activation of a transcription factor, FoxO (Forkhead transcription factors), which is responsible for stimulating the transcription of atrogenes, such as MuRF1 (muscle RING-finger protein-1) and Atrogin. The metformin treatment leads to a decreasing in tumour proliferation and an increase in tumour cell death by apoptosis, which leads to decrease on the production and release of a proteolytic factor, known as a proteolysis-inducing factor (PIF) or Walker factor (or PIF-like, WF). The decreasing in circulation levels of this proteolytic factor results in lower activation of protein breakdown in muscle cells. Also, the maintenance of circulating levels of insulin and IGF1 keep the Akt pathway working and maintenance of muscle mass.

Conclusion

Summing up our results and by other studies, the present chapter obtained some conclusions about metformin treatment, such as the effectiveness in decreasing Walker tumour cell proliferation and viability, both *in vivo* and *in vitro*. The metformin treatment increases key proteins such as AMPK, Bax, caspase 3, p53 in tumour tissue, so may have been the cause of increased death and reduced cell proliferation (Ki-67, PCNA labelled cells) by apoptosis (Figure 1). Moreover, the interesting point, probably related to the decreased tumour growth, is that metformin treatment reduced cachexia, as far as keeping the gastrocnemius muscle mass, likely concerned to the maintenance of the anabolic hormones balance and decreased expression of enzymes responsible for the process of muscle protein degradation (Figure 1). Even though the exact mechanism by which metformin act has not been described yet, metformin is still the first choice in the management of diabetes treatment because it has efficient action to decrease blood glucose levels with minor side effects; with well tolerance by patients and very low-cost. These reasons make metformin an important treatment with lower chances to be replaced. In the face of its use in diabetes treatment, metformin emerged as a promising drug in the oncology field with real potential, which has been evaluated over the last two decades. Our studies and by others have provided evidence that metformin can be in fact a good tool to understand better the basic molecular mechanisms related to tumour growth and survive. Further studies have to be conducted to better address the question of how metformin acts within the cell and what are its targets.

Acknowledgments

Our previous results were supported by FAPESP – Sao Paulo, Brazil (2017/02739-4).

REFERENCES

Alimova, Irina N., Bolin Liu, Zeying Fan, Susan M. Edgerton, Thomas Dillon, Stuart E. Lind, and Ann D. Thor. 2009. "Metformin Inhibits Breast Cancer Cell Growth, Colony Formation and Induces Cell Cycle Arrest *in vitro*." *Cell Cycle* 8 (6). Taylor & Francis: 909-15. doi:10.4161/cc.8.6.7933.

Argilés, Josep M., Silvia Busquets, Rodrigo Moore-Carrasco, Maite Figueras, Vanessa Almendro, and Francisco J. López-Soriano. 2007. "Targets in Clinical Oncology: The Metabolic Environment of the Patient." *Frontiers in Bioscience : A Journal and Virtual Library* 12 (January): 3024-51.

Bailey, Clifford J. 2017. "Metformin: Historical Overview." *Diabetologia* 60 (9): 1566-76. doi:10.1007/s00125-017-4318-z.

Balmanno, K., and S. J. Cook. 2009. "Tumour Cell Survival Signalling by the ERK1/2 Pathway." *Cell Death & Differentiation* 16 (3): 368-77. doi:10.1038/cdd.2008.148.

Chen, Yongqiang, Eileen McMillan-Ward, Jiming Kong, Sara J. Israels, and Spencer B. Gibson. 2007. "Mitochondrial Electron-Transport-Chain Inhibitors of Complexes I and II Induce Autophagic Cell Death Mediated by Reactive Oxygen Species." *Journal of Cell Science* 120 (Pt 23). The Company of Biologists Ltd: 4155-66. doi:10.1242/jcs.011163.

Choudry, Haroon A., Ming Pan, Anne M. Karinch, and Wiley W. Souba. 2006. "Branched-Chain Amino Acid-Enriched Nutritional Support in Surgical and Cancer Patients." *The Journal of Nutrition* 136 (1 Suppl): 314S-8S.

Coughlin, S. S., E. E. Calle, L. R. Teras, J. Petrelli, and M. J. Thun. 2004. "Diabetes Mellitus as a Predictor of Cancer Mortality in a Large Cohort of US Adults." *American Journal of Epidemiology* 159 (12): 1160-67. doi:10.1093/aje/kwh161.

Cruz, Bread, André Oliveira, Gislaine Ventrucci, and Maria Cristina Cintra Gomes-Marcondes. 2019. "A Leucine-Rich Diet Modulates the mTOR Cell Signalling Pathway in the Gastrocnemius Muscle under Different

Walker-256 Tumour Growth Conditions." *BMC Cancer* 19 (1). BioMed Central: 349. doi:10.1186/s12885-019-5448-0.

Decensi, Andrea, Matteo Puntoni, Pamela Goodwin, Massimiliano Cazzaniga, Alessandra Gennari, Bernardo Bonanni, and Sara Gandini. 2010. "Metformin and Cancer Risk in Diabetic Patients: A Systematic Review and Meta-Analysis." *Cancer Prevention Research* (Philadelphia, Pa.) 3 (11): 1451-61. doi:10.1158/1940-6207.CAPR-10-0157.

Denko, Nicholas C. 2008. "Hypoxia, HIF1 and Glucose Metabolism in the Solid Tumour." *Nature Reviews. Cancer* 8 (9): 705-13. doi:10.1038/nrc2468.

Dhillon, A. S., S. Hagan, O. Rath, and W. Kolch. 2007. "MAP Kinase Signalling Pathways in Cancer." *Oncogene* 26 (22): 3279-90. doi:10.1038/sj.onc.1210421.

Draznin, B. 2010. "Mitogenic Action of Insulin: Friend, Foe or 'frenemy'?" *Diabetologia* 53 (2). Springer-Verlag: 229-33. doi:10.1007/s00125-009-1558-6.

Edén, E., S. Edström, K. Bennegård, T. Scherstén, and K. Lundholm. 1984. "Glucose Flux in Relation to Energy Expenditure in Malnourished Patients with and without Cancer during Periods of Fasting and Feeding." *Cancer Research* 44 (4): 1718-24.

Evans, Josie M. M., Louise A. Donnelly, Alistair M. Emslie-Smith, Dario R. Alessi, and Andrew D. Morris. 2005. "Metformin and Reduced Risk of Cancer in Diabetic Patients." *BMJ* (Clinical Research Ed.) 330 (7503): 1304-5. doi:10.1136/bmj.38415.708634.F7.

Evans, William J., John E. Morley, Josep Argilés, Connie Bales, Vickie Baracos, Denis Guttridge, Aminah Jatoi, et al. 2008. "Cachexia: A New Definition." *Clinical Nutrition* (Edinburgh, Scotland) 27 (6): 793-99. doi:10.1016/j.clnu.2008.06.013.

Garcia, Daniel, and Reuben J. Shaw. 2017. "AMPK: Mechanisms of Cellular Energy Sensing and Restoration of Metabolic Balance." *Molecular Cell* 66 (6). NIH Public Access: 789-800. doi:10.1016/j.molcel.2017.05.032.

Gumucio, Jonathan P., and Christopher L. Mendias. 2013. "Atrogin-1, MuRF-1, and Sarcopenia." *Endocrine* 43 (1): 12-21. doi:10.1007/s12020-012-9751-7.

Gupta, K., G. Krishnaswamy, A. Karnad, and A. N. Peiris. 2002. "Insulin: A Novel Factor in Carcinogenesis." *The American Journal of the Medical Sciences* 323 (3). Elsevier: 140-45. doi:10.1097/00000441-200203000-00004.

Hammoudi, Naima, Kausar Begam Riaz Ahmed, Celia Garcia-Prieto, and Peng Huang. 2011. "Metabolic Alterations in Cancer Cells and Therapeutic Implications." *Chinese Journal of Cancer* 30 (8): 508-25. doi:10.5732/cjc.011.10267.

Hardie, D. Grahame, and Simon A. Hawley. 2001. "AMP-Activated Protein Kinase: The Energy Charge Hypothesis Revisited." *BioEssays* 23 (12): 1112-19. doi:10.1002/bies.10009.

Hasselgren, Per-Olof, Curtis Wray, and Joshua Mammen. 2002. "Molecular Regulation of Muscle Cachexia: It May Be More Than the Proteasome." *Biochemical and Biophysical Research Communications* 290 (1). Academic Press: 1-10. doi:10.1006/BBRC.2001.5849.

Hemkens, L. G., U. Grouven, R. Bender, C. Günster, S. Gutschmidt, G. W. Selke, and P. T. Sawicki. 2009. "Risk of Malignancies in Patients with Diabetes Treated with Human Insulin or Insulin Analogues: A Cohort Study." *Diabetologia* 52 (9): 1732-44. doi:10.1007/s00125-009-1418-4.

Hoshino, Rika, Yuji Chatani, Takao Yamori, Takashi Tsuruo, Hiroya Oka, Osamu Yoshida, Yutaka Shimada, et al. 1999. "Constitutive Activation of the 41-/43-kDa Mitogen-Activated Protein Kinase Signaling Pathway in Human Tumors." *Oncogene* 18 (3). Nature Publishing Group: 813-22. doi:10.1038/sj.onc.1202367.

Huang, Cai, Christoph H. Borchers, Michael D. Schaller, and Ken Jacobson. 2004. "Phosphorylation of Paxillin by p38MAPK Is Involved in the Neurite Extension of PC-12 Cells." *The Journal of Cell Biology* 164 (4). Rockefeller University Press: 593-602. doi:10.1083/jcb.200307081.

Inoki, Ken, Michael N. Corradetti, and Kun-Liang Guan. 2005. "Dysregulation of the TSC-mTOR Pathway in Human Disease." *Nature Genetics* 37 (1): 19-24. doi:10.1038/ng1494.

Jonasson, J. M., R. Ljung, M. Talbäck, B. Haglund, S. Gudbjörnsdòttir, and G. Steineck. 2009. "Insulin Glargine Use and Short-Term Incidence of Malignancies - a Population-Based Follow-up Study in Sweden." *Diabetologia* 52 (9): 1745-54. doi:10.1007/s00125-009-1444-2.

Kim, Eun Kyung, and Eui-Ju Choi. 2010. "Pathological Roles of MAPK Signaling Pathways in Human Diseases." *Biochimica et Biophysica Acta (BBA) - Molecular Basis of Disease* 1802 (4): 396-405. doi:10.1016/j.bbadis.2009.12.009.

Kourelis, Taxiarchis V., and Robert D. Siegel. 2012. "Metformin and Cancer: New Applications for an Old Drug." *Medical Oncology* 29 (2): 1314-27. doi:10.1007/s12032-011-9846-7.

Larsson, Susanna C., Christos S. Mantzoros, and Alicja Wolk. 2007. "Diabetes Mellitus and Risk of Breast Cancer: A Meta-Analysis." *International Journal of Cancer* 121 (4): 856-62. doi:10.1002/ijc.22717.

Libby, Gillian, Louise A. Donnelly, Peter T. Donnan, Dario R. Alessi, Andrew D. Morris, and Josie M. M. Evans. 2009. "New Users of Metformin Are at Low Risk of Incident Cancer: A Cohort Study among People with Type 2 Diabetes." *Diabetes Care* 32 (9): 1620-25. doi:10.2337/dc08-2175.

Ling, Sunbin, Qiaonan Shan, Peng Liu, Tingting Feng, Xuanyu Zhang, Penghui Xiang, Kangchen Chen, et al. 2017. "Metformin Ameliorates Arsenic Trioxide Hepatotoxicity via Inhibiting Mitochondrial Complex I." *Cell Death & Disease* 8 (11). Nature Publishing Group: e3159-e3159. doi:10.1038/cddis.2017.482.

Maclaine, Nicola J., and Ted R. Hupp. 2009. "The Regulation of p53 by Phosphorylation: A Model for How Distinct Signals Integrate into the p53 Pathway." *Aging* 1 (5): 490-502.

Malumbres, Marcos, and Mariano Barbacid. 2003. "RAS Oncogenes: The First 30 Years." *Nature Reviews Cancer* 3 (6): 459-65. doi:10.1038/nrc1097.

Manning, Brendan D., Andrew R. Tee, M. Nicole Logsdon, John Blenis, and Lewis C. Cantley. 2002. "Identification of the Tuberous Sclerosis Complex-2 Tumor Suppressor Gene Product Tuberin as a Target of the Phosphoinositide 3-Kinase/akt Pathway." *Molecular Cell* 10 (1): 151-62.

Marble, Alexander. 1934. "Diabetes and Cancer." *New England Journal of Medicine* 211 (8): 339-49. doi:10.1056/NEJM193408232110801.

McCarty, Mark F. 2004. "Chronic Activation of AMP-Activated Kinase as a Strategy for Slowing Aging." *Medical Hypotheses* 63 (2): 334-39. doi:10.1016/j.mehy.2004.01.043.

Mendes, Maria Carolina S., Gustavo D. Pimentel, Felipe O. Costa, and José B. C. Carvalheira. 2015. "Molecular and Neuroendocrine Mechanisms of Cancer Cachexia." *Journal of Endocrinology* 226 (3): R29-43. doi:10.1530/JOE-15-0170.

Menendez, Daniel, Alberto Inga, and Michael A. Resnick. 2009. "The Expanding Universe of p53 Targets." *Nature Reviews. Cancer* 9 (10). Nature Publishing Group: 724-37. doi:10.1038/nrc2730.

Oliveira, A. G., and M. C. C. Gomes-Marcondes. 2014. "Metformin Improves Carbohydrate Metabolism and Minimizes Walker Tumor Growth in Young Rats." *Biochemistry & Pharmacology: Open Access*. OMICS Group Inc., doi:10.4172/2167-0501.1000125.

Oliveira, André G., and Maria Cristina Cintra Gomes-Marcondes. 2016. "Metformin Treatment Modulates the Tumour-Induced Wasting Effects in Muscle Protein Metabolism Minimising the Cachexia in Tumour-Bearing Rats." *BMC Cancer* 16 (1): 418. doi:10.1186/s12885-016-2424-9.

Owen, M. R., E. Doran, and A. P. Halestrap. 2000. "Evidence That Metformin Exerts Its Anti-Diabetic Effects through Inhibition of Complex 1 of the Mitochondrial Respiratory Chain." *The Biochemical Journal* 348 Pt 3 (June): 607-14.

Pollak, Michael. 2008. "Insulin and Insulin-like Growth Factor Signalling in Neoplasia." *Nature Reviews Cancer* 8 (12): 915-28. doi:10.1038/nrc 2536.

Ruderman, N. B., J. M. Cacicedo, S. Itani, N. Yagihashi, A. K. Saha, J. M. Ye, K. Chen, et al. 2003. "Malonyl-CoA and AMP-Activated Protein Kinase (AMPK): Possible Links between Insulin Resistance in Muscle and Early Endothelial Cell Damage in Diabetes." *Biochemical Society Transactions* 31 (1): 202-6. doi:10.1042/bst0310202.

Sasaoka, T., D. W. Rose, B. H. Jhun, A. R. Saltiel, B. Draznin, and J. M. Olefsky. 1994. "Evidence for a Functional Role of Shc Proteins in Mitogenic Signaling Induced by Insulin, Insulin-like Growth Factor-1, and Epidermal Growth Factor." *The Journal of Biological Chemistry* 269 (18): 13689-94.

Shaul, Yoav D., and Rony Seger. 2007. "The MEK/ERK Cascade: From Signaling Specificity to Diverse Functions." *Biochimica et Biophysica Acta (BBA) - Molecular Cell Research* 1773 (8). Elsevier: 1213-26. doi: 10.1016/J.BBAMCR.2006.10.005.

Tisdale, Michael J. 2002. "Cachexia in Cancer Patients." *Nat. Rev. Cancer* 2 (11): 862-71. doi:10.1038/nrc927.

Tisdale, Michael J. 2009. "Mechanisms of Cancer Cachexia." *Physiol. Rev.* 89 (2): 381-410. doi:10.1152/physrev.00016.2008.

Vigneri, Paolo, Francesco Frasca, Laura Sciacca, Giuseppe Pandini, and Riccardo Vigneri. 2009. "Diabetes and Cancer." *Endocrine-Related Cancer* 16 (4): 1103-23. doi:10.1677/ERC-09-0087.

Warburg, O. 1956. "On Respiratory Impairment in Cancer Cells." *Science* (New York, N. Y.) 124 (3215): 269-70.

Wheaton, William W., Samuel E. Weinberg, Robert B. Hamanaka, Saul Soberanes, Lucas B. Sullivan, Elena Anso, Andrea Glasauer, et al. 2014. "Metformin Inhibits Mitochondrial Complex I of Cancer Cells to Reduce Tumorigenesis." *eLife* 3 (May). eLife Sciences Publications, Ltd: e02242. doi:10.7554/eLife.02242.

"World Health Organization - International Agency for Research on Cancer." 2014.

Wullschleger, Stephan, Robbie Loewith, and Michael N. Hall. 2006. "TOR Signaling in Growth and Metabolism." *Cell* 124 (3): 471-84. doi:10.1016/j.cell.2006.01.016.

Yano, Claudia L. C. L., Gislaine Ventrucci, William N. W. N. Field, M. J. Michael J. Tisdale, and M. C. Maria Cristina C. Gomes-Marcondes. 2008. "Metabolic and Morphological Alterations Induced by Proteolysis-Inducing Factor from Walker Tumour-Bearing Rats in C2C12 Myotubes." *BMC Cancer* 8: 24. doi:10.1186/1471-2407-8-24.

Zhou, G., R. Myers, Y. Li, Y. Chen, X. Shen, J. Fenyk-Melody, M. Wu, et al. 2001. "Role of AMP-Activated Protein Kinase in Mechanism of Metformin Action." *The Journal of Clinical Investigation* 108 (8): 1167-74. doi:10.1172/JCI13505.

In: The Pharmacological Guide to Metformin ISBN: 978-1-53616-634-7
Editor: Natalie Lajoie

Chapter 5

THE THERAPEUTIC POTENTIAL OF METFORMIN IN BREAST AND OVARIAN CANCER

Isabella dos Santos Guimarães[1], Diandra Zipinotti dos Santos[2], Paulo C. M. Lyra-Junior[2], Ian Victor Silva[2,3,4] and Leticia Batista Azevedo Rangel[2,4,5,*]

[1]Clinical Research Division,
Brazilian National Cancer Institute (INCA), Rio de Janeiro, RJ, Brazil
[2]Biotechnology Program/RENORBIO, Health Sciences Center,
Federal University of Espirito Santo, Vitoria, ES, Brazil
[3]Department of Morphology, Health Sciences Center,
Federal University of Espirito Santo, Vitoria, ES, Brazil
[4]Biochemistry Program, Health Sciences Center,
Federal University of Espirito Santo, Vitoria, ES, Brazil
[5]Department of Pharmaceutical Sciences, Health Sciences Center,
Federal University of Espirito Santo, Vitoria, ES, Brazil

[*] Corresponding Author's E-mail: lbarangel@yahoo.com.

ABSTRACT

Metformin is the first-line drug used in the treatment of type 2 diabetes with more than six decades of clinical use. It is a relatively safe drug, with known pharmacokinetics and a favorable safety profile. In recent years, in addition to hypoglycemic effect, several epidemiological studies have highlighted the therapeutic potential of metformin in the context of other diseases, especially in breast and ovarian cancers; thus, demonstrating its beneficial actions on survival outcome and prevention of these highly lethal malignancies.

Recent and robust data support the fact that metformin would provide invaluable benefits when administrated along with conventional adjuvant therapy against the breast and ovarian cancers. Metformin impairs cellular metabolism, suppresses oncogenic signaling pathways, inhibits cell growth and metastasis, as well as reverses chemoresistance. Thus, improving treatment efficacy and allowing the prescription of lower doses of highly toxic antineoplastic agents.

Altogether, the aforementioned information points out metformin as an anticancer agent. Of clinical interest, it arises as a therapeutic possibility in the fight against poor prognosis diseases, such as triple-negative breast cancer and high grade serous ovarian cancer. In the present chapter, we summarized the molecular mechanisms underlying the anticancer effects of metformin in breast and ovarian cancers. Furthermore, we discuss the potential anticancer applications of metformin-based combinatorial chemotherapy.

INTRODUCTION

Metformin (1,1-dimethylbiguanide) is the first-line pharmacological therapy for type 2 diabetes mellitus (T2DM) in almost all guidelines and recommendations worldwide due to its excellent hypoglycemic and cardiovascular protective effects (Clifford J. Bailey 2017; Holman et al. 2008). It is a medication derived from French Lilac or goat's rue (*Galega officinalis*), a traditional herbal medicine for the treatment of polyuria in medieval Europe (C. J. Bailey and Day 1989). *Galega officinalis* was rich in guanidine responsible for hypoglycemic proprieties and this led to the development of three main biguanides: buformin, phenformin and metformin.

Phenformin and buformin was used to treat T2DM but were discontinued in the late of 1970s as a result of high risk of lactic acidosis and increased cardiac mortality (Luft, Schmolling, and Eggstein 1978; Clifford J. Bailey 2017).

Metformin, the less potent glucose-lowering biguanides, was being used in the 1940s to treat influenza and malaria exhibiting excellent tolerability, safety profile and reduce blood glucose levels. This drug was approved for use in the 1970s in Europe and in the 1990s in the United States in the management of T2DM (Clifford J. Bailey 2017). It is estimated that there are over 120 million annual prescriptions for metformin worldwide (Ben Sahra et al. 2010). Metformin is also prescribed for treating polycystic ovarian syndrome (PCOS), metabolic syndrome and gestational diabetes mellitus (Lashen 2010; Finneran and Landon 2018).

The most frequent adverse events of metformin at therapeutic doses are metallic taste and mild-to-moderate gastrointestinal discomfort (diarrhea, nausea, vomiting, bloating and abdominal pain), usually transient which can be alleviate by gradual dose escalation or divide doses with meals (Dujic et al. 2015). Slow-release formulations are associated with fewer gastrointestinal symptoms (Davidson and Howlett 2004). Gastrointestinal intolerance is related to high concentration of metformin in the intestinal lumen after oral administration of the drug and occurs in approximately 5-10% of patients leading to premature discontinuation of treatment even with the lowest dose (C. J. Bailey, Wilcock, and Scarpello 2008; Dawed et al. 2019). The precise mechanism of metformin gastrointestinal side effects have not been fully elucidated. Metformin may exert these effects by modulation intestinal levels of histamine and serotonin, regulation of the incretin axis by increasing glucagon-like peptide-1 (GLP-1) concentrations, bile-salt malabsorption, along with changes in the gut microbiota (Cubeddu et al. 2000; Carter et al. 2003; Mannucci et al. 2001; Forslund et al. 2015; Dujic et al. 2015).

Its most serious toxicity, lactic acidosis, is rare (<10 cases per 100,000 patient-years) and is not greater than with non-metformin therapies. Metformin-associated lactic acidosis (MALA), although rare, is more common in elderly patients (older than 80 years) and in those with impaired cardiac, hepatic and/or renal function (Misbin et al. 1998; DeFronzo et al. 2016). Of note, metformin is rarely associated with hypoglycemia (Barnett 2010) while the continuous use of this medication is related with vitamin B12 deficiency and anemia in a dose-dependent manner (Aroda et al. 2016). Thus, vitamin B12 monitoring for metformin-treated individuals and appropriate supplementation should be implemented (Guimaraes et al. 2015; Q. Liu et al. 2014).

Recently, in addition to hypoglycemic effect, the use of metformin received tremendous recognition in cancer research when retrospective studies suggested that metformin may also have beneficial effects for reducing cancer risk. Moreover, the safety, low cost, known pharmacokinetics and manageable toxicities of metformin have made this biguanide an attractive candidate for drug repurposing in cancer. In this chapter, we provide a comprehensive overview of the anticancer actions and mechanisms of metformin, given more focus on the fight against breast and ovarian cancer.

The Pharmacokinetics of Metformin

Metformin is a hydrophilic compound that exists as cationic species at physiological pH with low membrane permeability. Consequently, several transporters are required for the intracellular internalization of metformin in different tissues affecting pharmacokinetics (oral absorption, plasma concentration, distribution and elimination) and drug responses. These include organic cation transporters (OCT) 1–3, the multidrug and toxin extrusion protein (MATE) 1-2, the plasma monoamine transporter (PMAT), serotonin transporter (SERT) and high-affinity choline transporter (CHT) (Tucker et al. 1981; Graham et al. 2011; T. Han et al. 2015; S. W. Yee et al. 2015; Dujic et al. 2016; Gormsen et al. 2016).

Interestingly, recent data from pharmacogenetic studies on metformin revealed that the variation of transporter genes is one of the major factors affecting therapeutic response to metformin (Shikata et al. 2007; Shu et al. 2008; Christensen et al. 2011; Dujic et al. 2016; Florez 2017).

Oral bioavailability of metformin has been estimated at 50-60%. The absorption occurs predominantly in the small intestine (duodenum and jejunum), with approximately 30% dose recovery of unchanged metformin from faeces (McCreight, Bailey, and Pearson 2016). Once absorbed, metformin is mainly distributed in the liver, intestines, kidneys, salivary glands and bladder with only slow accumulation in muscle (Dunn and Peters 1995; Gormsen et al. 2016). Several studies have reported that metformin is excreted into urine unchanged (not metabolized) principally by active tubular secretion in the kidney (Dunn and Peters 1995; Scheen 1996). Metformin does not bind to plasma proteins; it has an elimination half-life of approximately 5–6 hours in patients with physiological renal function and effects of this medication continue for 8–12 hours (Graham et al. 2011; Gong et al. 2016).

After a single dose, peak plasma concentrations are achieved after approximately 2-3 hours in diabetic patients. The plasma concentration of metformin ranges from 6 μM up to approximately 18 μM. When tested in clinical trials, plasma levels never exceed ~30 μM (C. J. Bailey and Day 1989; Tucker et al. 1981; Graham et al. 2011). Recently, experiments in animal models and positron emission tomography (PET) in humans demonstrate that much greater metformin concentrations are achieved in the liver than in the plasma (estimated concentration in the human liver is about 50–100 μM) (Jensen et al. 2016; Gormsen et al. 2016). Once in the cell, metformin concentrates in the cytoplasm and then in the mitochondrial matrix, due to their positive charge (Owen, Doran, and Halestrap 2000). This physical phenomenon results in an intramitochondrial concentration of metformin about 100-fold higher than the external medium in cultured cells and 1000-fold in the mitochondria matrix relative to the serum in *in vivo* experiments (Bridges et al. 2014).

Metformin as an Anti-Cancer Agent: Potential Mechanisms of Action

The precise molecular mechanism of metformin is considered an active area of investigation, because it remains partially understood and controversial. Compelling evidence has revealed that anti-tumor activity of metformin can be divided into two different but not exclusive mechanisms: an indirect effect (through lowering insulin and blood glucose levels) and direct mechanism acting on cancer cells. Metformin´s indirect or direct action involves AMPK-dependent and -independent mechanisms (Figure 1).

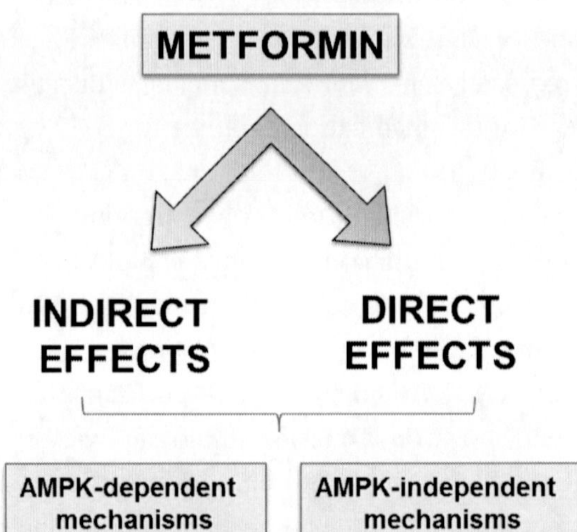

Figure 1. Metformin anti-cancer effects.

Indirect Effects

Associations between diabetes and cancer have been established by several meta-analyses (Giovannucci et al. 2010; Tsilidis et al. 2015; Cignarelli et al. 2018; Vigneri and Sciacca 2018). It should be emphasized, that is not known whether the observed associations were due to diabetes exclusively or other prognostic factors. However, it is plausible that metformin can exert an antitumor effect through its action on hyperglycemia and hyperinsulinemia. T2DM patients show resistance to insulin action and, at least in the early stages of the disease, excess insulin production. Several studies have suggested that insulin might have a mitogenic effect through insulin-like growth factors receptor (IGF-1R) and insulin receptor (IR) (Pollak 2012). IGF-1R phosphorylation leads to activation of two major growth factor signaling pathways: phosphatidylinositol-3cinase (PI3K)/Protein kinase b (AKT) and mitogen-activated protein kinase (MEK)/extracellular signal–regulated kinase (ERK), which have important functions in growth, metabolism, survival and cell proliferation (Chitnis et al. 2008; Annenkov 2009). In fact, they are widely studied intracellular signaling pathways and most frequently altered in ovarian and breast cancer (T. W. Miller et al. 2010; Huang et al. 2011; Mabuchi et al. 2015). Insulin-like growth factors 1 is frequently overexpressed in cancer cells, including breast cancers (Peyrat et al. 1993; Christopoulos, Msaouel, and Koutsilieris 2015; D. Yee 2018). Preclinical studies also indicate that inhibition of insulin-like growth factors/insulin pathway leads to a reduction of breast cancer cells proliferation and tumor metastasis (H.-J. Kim et al. 2007; Sachdev et al. 2004, 2003).

Moreover, not only hyperglycemia but an excessive supply of glucose in the bloodstream may create an ideal environment for cancer progression. Also, hyperglycemia can induce the emergence of oxidative stress, which may promote carcinogenesis (Johnson et al. 2012; Ryu, Park, and Scherer 2014; Hou et al. 2017).

Metformin is certainly an effective glucose-lowering agent and this medication appears to have different actions on several pathways that are individually modest, but collectively relevant for hypoglycemic effects. It is widely accepted that metformin acts on the liver, gut, adipose tissue and striated muscle to lower blood glucose levels in diabetic patients. The primary molecular target of metformin is believed to be mitochondria, where it inhibited Complex I (NADH:ubiquinone oxidoreductase) of the electron transport chain leading to decrease in ATP synthesis and elevated cellular (AMP)/ (ATP) and (ADP)/(ATP) ratio (El-Mir et al. 2000; Owen,

Doran, and Halestrap 2000). The induction of this cellular stress is responsible for the activation of the 5′ adenosine monophosphate-activated protein kinase (AMPK) via phosphorylation by the upstream kinases liver kinase B1 (LKB1) or Ca2+/calmodulin-dependent protein kinase kinase β (CaMKK2) (Hawley et al. 2005; Woods et al. 2003). AMPK is the key regulator of cellular energy homeostasis in eukaryotes, coordinating a metabolic switch from anabolism towards catabolism under energy deprivation. Activation of hepatic AMPK leads to inhibition of hepatic lipogenesis and gluconeogenesis (Kalender et al. 2010; Hardie, Ross, and Hawley 2012; Krishan, Richardson, and Sahni 2015).

Subsequently, studies proposed decrease in cellular energy state involving AMPK-independent mechanisms. Foretz and colleagues demonstrated that metformin inhibits hepatic gluconeogenesis independent of AMPK using liver-specific AMPK knockout mice (Foretz et al. 2010). Additionally, metformin can inhibit hepatic glucose production through AMP-mediated inhibition of adenylate cyclase, reducing the amount of glucagon-increased cyclic adenosine monophosphate (cAMP), which in turn greatly reduces glucagon-mediated gluconeogenesis (R. A. Miller et al. 2013; Baur and Birnbaum 2014). Another possible mechanism of acute inhibition of glucose production by metformin was due to inhibition of fructose-1,6-bisphosphatase (FBP1) by AMP (Hunter et al. 2018; Alshawi and Agius 2019).

Likewise, recent studies by Madiraju et al. have shown that metformin inhibits mitochondrial glycerophosphate dehydrogenase (mGPDH) and thereby attenuated transfer of Nicotinamide Adenine Dinucleotide (NADH) from the cytoplasm into mitochondria. Consequently, the cellular redox state is altered and decreased conversion of lactate and glycerol to glucose reducing hepatic gluconeogenesis (Madiraju et al. 2014). Recent reports demonstrate that redox-modulation is a key element required to metformin-mediated inhibition of gluconeogenesis *in vivo*, independently of gluconeogenic enzyme protein expression (Madiraju et al. 2018). Notably, these effects require lower doses (micromolar concentrations) than are needed to affect AMPK (supra-pharmacological concentrations). Nevertheless, recent advances demonstrate that other actions on the gastrointestinal tract and even a gut–brain–liver axis appear to account for most of the glucose-lowering effect of metformin at therapeutic doses (Buse et al. 2015).

Direct Effects

Accumulating evidence suggests that direct anticancer action of metformin is mainly centered on its ability to activate AMPK pathway that affects several cancer cellular processes. Activation of AMPK induced by metformin results in the inhibition of mammalian target of rapamycin complex 1 (mTORC1). mTORC1, one of the two functional mTOR complexes, regulates the protein synthesis by directly phosphorylating two important targets involved in this process: 40S ribosomal protein S6 kinase (p70S6K) and 4E-binding protein 1 (4E-BP1) (Hara et al. 1998).

AMPK inhibits mTOR signaling through two distinct mechanisms: (i) phosphorylation of tuberous sclerosis complex 2 (TSC2), a subunit of the TSC1/TSC2 complex. This complex converts the mTOR activator Ras Homolog Enriched in Brain (Rheb) to its inactive GDP-bound form; (ii) direct phosphorylation of Raptor, a subunit of the mTORC1 complex (K. Inoki 2003; Ken Inoki, Corradetti, and Guan 2005; Gwinn et al. 2008). Interestingly, metformin inhibits the mTOR activity in an alternative mechanism independent of AMPK and TSC1/2 through recombinant activating gene (RAG) GTPase family protein inhibition (Kalender et al. 2010). Inactivation of RAG GTPases leads to dissociation of mTORC1 from Rheb. Additionally, metformin also inhibits mTOR independently of AMPK by up regulating the expression of Regulated in Development and DNA Damage responses (REDD1) resulting in cell cycle arrest (Ben Sahra et al. 2011).

Other widely accept effects of AMPK activation by metformin include phosphorylation and inactivation of several enzymes involved in lipid and cholesterol synthesis, such as acetyl-CoA carboxylase (ACC), fatty acid synthase (FAS) and 3-hydroxy-3-methyl-glutaryl-CoA (HMG-CoA) reductase (Lettieri Barbato et al. 2014). Furthermore, AMPK/mTOR suppresses the transcription factors sterol regulatory element-binding proteins (SREBPs) strongly involved in the control of lipid biogenesis (Yu Li et al. 2011; Y. Han et al. 2019). As they rapidly divide and proliferate, cancer cells require an increased rate of fatty acid synthesis to produce new cell membranes (E. Currie et al. 2013).

EFFECTS OF METFORMIN IN BREAST AND OVARIAN CANCER: PRECLINICAL STUDIES

Since numerous epidemiological studies associated the use of metformin with a lower risk of cancer incidence, this biguanide has received increasing interest as a potential anticancer agent. The antitumor effects of metformin have been extensively investigated in preclinical models, especially in breast and ovarian cancer.

Breast cancer is the most occurring cancer in women and the second cause of cancer-related deaths in the world, where 2,088,849 new cases of the disease and 626,679 related deaths yearly (Ferlay et al. 2018). Despite the notorious improvement in breast cancer treatment, the mortality rates are still impressive. At least partially, this is because breast cancer comprises multiple diseases which subtypes respond differently to treatment and are associated to distinct clinical outcomes (Perou et al. 2000; Sorlie et al. 2001). Estrogen receptor (ER), progesterone receptor (PR) and human epidermal growth factor receptor-2 (HER2) are the most relevant clinical biomarkers that are widely used in stratifying breast cancer with prognostic and therapeutic implications. Breast cancer is classified according to the immunohistochemical in: luminal A (ER+ and/or PR+ and HER2 -), luminal B (ER+, PR+ and HER2+), HER2 enriched (ER-, PR-, HER2+) and triple-negative breast cancer (TNBC) (ER-, PR-, HER2-) (Torregrosa et al. 1997; Kakar 2000; Hammond et al. 2010)

As cited before, there is an association between diabetes and cancer. Epidemiological studies indicate that women with T2DM are more likely to develop breast cancer and diabetes is linked to poor overall survival and disease-free survival (Boyle et al. 2012; X.-B. Zhao and Ren 2016). Importantly, dyslipidemia, hyperinsulinemia and hyperglycemia, the three main metabolic disturbances that occur in diabetes, might contribute to the development of an oncogenic phenotype of the breast tissue. In this scenario, the antitumor effects of metformin have been extensively investigated in preclinical models.

Currently, series of scientific works outline the anti-proliferative properties of metformin in different subtypes of breast cancer cells *in vitro* and *in vivo.* Alimova and colleagues (2009) have shown that metformin can inhibit the MAPK and PI3K/AKT/mTOR pathway in ER-positive and ER-negative, HER2 overexpressing and HER2 normal expressing breast cancer cell lines (Alimova et al. 2009).

Within subtypes of breast cancer, TNBC represents a heterogeneous group with distinct molecular signature and biological pathways, being the most aggressive and invasive subtype with unfavorable prognosis (Thike et al. 2010; Patel et al. 2018). TNBC accounts approximately for 15-20% of all new breast cancer diagnoses annually (Gierach, Burke, and Anderson 2010). Patients with TNBC do not benefit from the use of endocrine and target therapies. The lack of effective therapeutic options for this subtype, still present a challenge for clinicians who often have to resort to highly unspecific cytotoxic therapies (Addington and Freimer 2016). This makes it desirable to identify drugs which synergize with chemotherapeutics and potentially reduce the dose of them that is necessary to treat patients.

Some authors have found that metformin decreases cell proliferation and induces apoptosis also in TNBC cells (B. Liu et al. 2009; Deng et al. 2012; Zordoky et al. 2014). In addition, in TNBC mouse xenograft model, metformin significantly suppresses tumor proliferation and increases life span (Liu et al., 2009). Another study in TNBC cells showed that metformin reduced growth and inhibited migration and invasion by AMPK-induced glycogen synthase kinase 3β (GSK3β) and sirtuin1 (SIRT1) affecting c-Myc expression (Gollavilli et al. 2015). *In vitro* treatment in TNBC cells with metformin resulted in downregulation of genes involved in fatty acid and cholesterol biosynthesis pathways. As well, mammosphere formation was reduced by metformin-stimulated increase in miR-193 (Wahdan-Alaswad et al. 2014). Additionally, Sharma and colleagues (2019) also showed that metformin has anti-tumorigenic activity by lowering of cholesterol level in TNBC cells (Sharma et al. 2019). Furthermore, metformin treatment suppressed expressions of anti-apoptotic genes and mesenchymal genes with simultaneous enhancement of apoptotic and epithelial genes expressions (Sharma et al. 2019).

Some studies showed that metformin directly influences proliferation and apoptosis of luminal A breast cancer cell line, through AMPK activation and PI3K/AKT/mTOR inhibition (Zakikhani et al. 2010; Dowling et al. 2015). Queiroz and colleagues (2014) showed that ER positive cells treated with metformin suffered cell cycle arrest, cell apoptosis and increased necrosis. These effects were associated with an increase in FOXO3a activity via AMPK activation, decreased the activation of the IR/PI3K/AKT/mTOR and IR/MEK/ERK1/2 pathways and decrease in the expression of insulin receptor (Queiroz et al. 2014).

Evidence from *in vitro* studies also suggests that metformin may represent a strategy for hormone-dependent breast carcinoma treatment. Metformin treatment repressed the expression and transcriptional activity of ERα. Interestingly, metformin exhibited a superior antiproliferative effect by inhibiting ERα signaling than tamoxifen in ERα-positive MCF-7, TR MCF-7 and MDA-MB-361 cells (Kim et al. 2016). In addition, metformin inhibited aromatase expression in primary human breast adipose stromal cells, through the inhibition of CREB regulated transcription co-activator 2 (CRTC2) mediated by AMPK (Samarajeewa et al. 2011).

HER2 is overexpressed in around 30% of breast cancers and activates several signaling pathways implicated in breast cancer progression (Harbeck and Gnant 2017). In preclinical study on breast cancer, metformin treatment resulted in reduced tumor development and size in transgenic mice carrying HER2 oncogene when administered along with drinking water (Anisimov et al. 2005). Metformin was reported to downregulate HER2 oncogene overexpression in human breast cancer cells through a direct and AMPK-independent inhibition of mTOR. Ectopical overexpression of the HER2 oncogene significantly enhanced metformin-induced growth inhibition (Vazquez-Martin et al. 2011). Interestingly, metformin notably blocked HER2 tyrosine kinase activity at low concentrations.

Moving to another perspective, ovarian cancer is the most lethal gynecologic malignancy and most of the cases are detected as advanced and metastatic disease, which hinders effective treatment in the clinic (Ferlay et al. 2018). Similar to what was observed for breast cancer, several studies suggest that the aggressive and poor outcome of ovarian cancer may be resulted from a constant state of hyperinsulinemia, elevated IGF1, increased estrogen levels and pro-inflammatory mediators (Howe et al. 2013). In fact, IGF-I and –II levels have been associated with decreased survival in epithelial ovarian cancer (Sayer et al. 2005; Yu 2000). The standard management of advanced ovarian cancer includes cytoreductive surgery followed by platinum-taxane based chemotherapy. However, subsequent drug resistance minimizes the effectiveness of chemotherapy drugs (du Bois 2005; Chang et al. 2013). Therefore, it is critical to develop more effective treatment approaches for ovarian cancer.

A recent experimental study by Huo et al. revealed that metformin has inhibitory effect on ovarian cancer cells SKOV3 and A2780. Metformin, used in different concentrations (2, 5, 10, 20 mM), reduced cell proliferation, decreased cell clone formation and induced apoptosis, showing a significant dose-dependence actions (Huo et al. 2017). In addition, metformin inhibits the ability of high-metastatic potential human ovarian cancer cell lines (SKOV3 and HO8910-PM) to adhere, invade and migrate *in vitro*. Metformin treatment also reduces hepatic, intestinal and lung metastases *in vivo* (Wu et al. 2012).

It has been stablish that omental adipocytes can attract the ovarian cancer cells to omentum and promote cancer growth the by secreting adipokines, cytokines and growth factors (Nieman et al. 2011). Tebbe and colleagues showed that metformin treatment modulated the ovarian cancer promoting effects of adipocytes and found that ID8 ovarian cancer cells treated with low concentrations of the antidiabetic drug (2-4 mM) had adipocyte-induced migration/invasion attenuated. Likewise, data suggest metformin exerts its effects through the regulation of key adipogenesis regulating transcription factors (CEBPα, CEBPß, and SREBP1) and AMPK activity (Tebbe et al. 2014). Taken together, these findings corroborate the hypothesis that metformin have a considerable influence on lipid pathway through activating AMPK, which lead to a possible beneficial use of metformin in hindering the initial step of ovarian cancer metastasis.

It's important to emphasize that although the antiproliferative effects of metformin have been extensively studied in cancer preclinical models, few studies have assessed the expression of OCT1 in human tumors. The intracellular uptake of metformin is enhanced in cancer cells with up-regulated OCT1 transporters and ovarian cancer cells with a downregulated OCT1 are resistant to metformin toxicity (Segal et al. 2011). It is noteworthy that assessment of metformin transporters expression patterns in tumor tissue will be critical to predict biguanide responses.

Metformin in Combination Therapies

The emergence of multidrug/chemotherapy resistant cells in breast and ovarian cancer has been challenging and new therapeutic approaches to overcome these impasses are necessary. There are now enough compelling evidences to suggest the role of metformin to re-sensitizes cancer cells and to cooperate with some anticancer drugs to improve efficacy through several cellular pathways. Additionally, it could reduce the concentration of anticancer agents in dosage and avoid unwanted toxic effects. A compilation of several reports showing synergistic or enhanced effects when metformin is used in combination with chemotherapeutic drugs and with targeted therapies is presented on Table 1.

Development of resistance constitutes a significant challenge to effective first-line platinum-taxane based therapy for ovarian cancer patients. The results of several studies have revealed that metformin inhibit cellular growth and enhances the sensitivity to cisplatin and paclitaxel in resistant ovarian cancer cells (Table 1). Several platinum-resistant ovarian cancer cell lines (A2780, CP70, C200, OV202, OVCAR3, SKVO3ip, PE01 and PE04) had their proliferation significantly reduced by the treatment of metformin (5–20 mM) in time- and dose-dependent manner (Ramandeep Rattan et al. 2011). Animal studies verified the synergistic effect of metformin with cisplatin in ovarian cancer. In an *in vivo* study, Rattan and colleagues discovered that metformin treatment inhibited the proliferation, metastasis and angiogenesis of ovarian tumors in xenograft models of A2780 ovarian cancer cells and significantly potentiated cisplatin-induced cytotoxicity by inhibiting tumor growth compared with treatment by either of the drugs alone (Ramandeep Rattan et al. 2011).

Also, metformin enhanced the anti-tumor effect of cisplatin and inhibit epithelial–mesenchymal transition (EMT) in epithelial ovarian cancer *in vitro* and *in vivo*. The co-treatment with metformin and cisplatin increased the level of pro-apoptotic proteins and decreased the level of anti-apoptotic proteins in ovarian cancer cells. Additionally, metformin and cisplatin co-treatment significantly inhibited ovarian cancer growth in xenograft mouse models (Zheng et al. 2018). Metformin downregulated Axl and Tyro3 receptor tyrosine kinases (RTKs) expression and subsequently deactivated ERK1/2 and STAT3, resulting in its anti-proliferative activity in both parental and cisplatin/taxol-resistant ovarian cancer cells (KIM, LEE, and LEE 2015). Another recent study indicated that metformin combined with cisplatin inhibited cell viability and induced apoptosis in human ovarian cancer cells by inactivating ERK 1/2 (Dang et al. 2017). Pre-treatment of SKOV3ip1 xenograft mice with metformin significantly reduced tumor formation, and co-treatment of metformin and paclitaxel reduced tumor weight by 60% compared with controls in *LSL*-K-ras$^{G12D/+}$PTEN$^{loxP/loxP}$ genetic mouse model (Lengyel et al. 2015).

Preclinical data has demonstrated that pretreatment and maintenance of micromolar concentrations of metformin in conjunction with carboplatin caused a synergistic enhancement in cytotoxicity of A2780 and SKOV3 ovarian cancer cell lines.

Table 1. Preclinical studies of metformin use in combination with other agents in cancer therapy

Type of cancer	Metformin in combination with	Experimental settings	Observed Results	Reference
Ovarian cancer	Cisplatin	HO-8910 cells	Inhibited cell viability and induces apoptosis by inactivating ERK ½	(Dang et al. 2017)
	Cisplatin	Three high-grade serous/endometrioid patient-derived xenografts	Increased cisplatin antitumor activity	(Ricci et al. 2019)
	Cisplatin	A2780-induced tumors in nude mice	Synergistically inhibited cell proliferation, angiogenesis	(R. Rattan et al. 2011)
	Cisplatin	SKOV3 and Hey cell lines Xenograft mouse models	Metformin effectively boosted the inhibitory effects of cisplatin on tumor proliferation and metastasis	(Zheng et al. 2018)
	Cisplatin + SB203580	SKOV3/DDP cisplatin-resistant cells of human ovarian serous cystadenocarcinoma	Improved sensitivity of cells to cisplatin treatment	(Lengyel et al. 2015)
	Cisplatin Paclitaxel	A2780, A2780-PR, ACRP cell lines	Improved sensitivity of cells to cisplatin and paclitaxel treatment; Reduction of inflammatory cytokines and activation of the NF-κB pathway with metformin in doses allowed for clinical use	(dos Santos Guimarães et al. 2018)
	Carboplatin Doxorubicin Paclitaxel	A2780 and SKOV3 cells	Enhanced the reduction in cell viability; Metformin overcame resistance to carboplatin	(R. Erices et al. 2013)
	Olaparib	UWB1.289, UWB1.289.BRCA, SKOV3, OVCAR5, A2780 and C200 cancer cells; Xenograft mouse models using SKOV3 and A2780	Metformin enhanced anti-proliferative activity in BRCA mutant and in BRCA intact cancer cells *in vitro* and *in vivo*	(Hijaz et al. 2016)

Type of cancer	Metformin in combination with	Experimental settings	Observed Results	Reference
Ovarian cancer	Paclitaxel	SKOV3ip1, HeyA8 and K-ras/PTEN ovarian cancer cells; K-ras/PTEN Genetic Mouse Model	Increased sensitivity to paclitaxel; Reduced tumor growth;	(Lengyel et al. 2015)
	Ciplatin + Paclitaxel Carboplatin + Docetaxel Nedaplatin + Docetaxel	Primary cell culture with omental metastatic (OM) cells	Metformin increased drug sensitivity in cells in a dose-dependent manner; Increased platinum drug-induced apoptosis	(Liu et al. 2018)
	Paclitaxel Everolimus	ES-2 and A2780 cell lines	Metformin or everolimus exerts synergistic effect with paclitaxel against epithelial ovarian cancer	(Tessarollo et al. 2015)
Breast cancer	Paclitaxel	MCF-7 cell line	Synergistic decrease in cell proliferation; increase in apoptosis, AMPK activation and G2/M phase cell cycle arrest	(Rocha et al. 2011)
	Doxorubicin	multi-drug-resistant breast cancer cells-MCF7/ADR.	Metformin enhanced the cytotoxicity of doxorubicin through attenuating hypoxic stress *in vitro* and *in vivo*	(Ying Li et al. 2019)
	Cisplatin Paclitaxel Doxorubicin	MDA-MB-231, HCC-70, HCC-1937, MCF-7 cells	Metformin enhanced the cytotoxicity of paclitaxel and doxorubicin in TNBC	(Guimaraes et al. 2015)
	Doxorubicin Paclitaxel Carboplatin	Mouse xenografts using MCF10A ER-Src, BT-474 and MDA-MB-231 cells.	Prolonged relapse and suppressed tumor growth; Metformin reduced the dose of doxorubicin necessary to exert the same effect	Iliopoulos et al. 2011.
	Everolimus	MCF-7 and MDA-MB-231 cells	Additive inhibitory effects on proliferation, colony formation, mTOR signaling activity and mitochondrial respiration.	(Ariaans et al. 2017)
	Everolimus	HCC1428, MDA-MB-468, and BT549 SCID mice injected with HCC1428	synergistic inhibition of cell growth and breast CSC	(Y. Wang et al. 2015)

Table 1. (Continued)

Type of cancer	Metformin in combination with	Experimental settings	Observed Results	Reference
Breast cancer	Carboplatin Doxorubicin Paclitaxel Carboplatin + Everolimus	MCF-7, T47-D, BT-474, and MDA-MB-231 cell lines	synergistic inhibition of cell proliferation and G1-phase cell cycle arrest	Liu H et al. 2012
	Rapamycin	MCF-7, T47-D, Hs-578T and MDA-MB-231 cell lines	Improvement of apoptosis and proliferation inhibition through PI3K/AKT/mTOR pathways than monotherapy	(Zakikhani et al. 2010)
	5-fluorouracil, Epirubicin, Cyclophosphamide	MDA-MB-468, MDA-MB-231, HCC1937, MCF-7, SKBR3, and T47D cell lines	Metformin synergizes cyclophosphamide in an AMPK-independent mechanism in breast CSCs	(Soo et al. 2015)
	Trastuzumab	SKBR3, MDA-MB-468, MDA-MB-231, SKBR3 TzbR, JIMT-1 cell lines	synergistic reduction of CSCs self-renewal and proliferation	(Vazquez-Martin et al. 2011)
	Trastuzumab	BT-474, SKBR3 and derived trastuzumab-resistant sublines BT474-HR20 (HR20) and SKBR3-pool2	synergistic inhibition of proliferation, clonogenicity and ErbB2/IGF-1R interactions through inactivating Src kinase and/or PI3K/Akt pathway	(B. Liu et al. 2011)
	Erlotinib	17 human breast cancer cell lines and MDA-MB-468 cell xenograft model.	synergistic reduction of mammosphere formation; Synergistic induction of apoptosis, cell death, and tumor regression through reduction of EGFR and AKT signaling	(Lau et al. 2014)

Similarly, in ovarian cancer primary cultures derived from the ascites of patients with advanced disease, micromolar doses of metformin improved the cytotoxic response to carboplatin but not paclitaxel or doxorubicin (Rafaela Erices et al. 2013). Consistent with that, our group has reported a significant anticancer activity of metformin at a micromolar concentration in combination with everolimus and paclitaxel in ovarian cancer cells *in vitro* (Tessarollo et al. 2015).

Our very recent study also showed that metformin decreased cellular proliferation of platinum and taxane-resistant ovarian cancer cells and reduced mRNA of inflammatory cytokines and NFκB signaling pathway (dos Santos Guimarães et al. 2018) The effect of metformin as an adjuvant to cisplatin-paclitaxel treatment during the acquirement of chemoresistance has been later explored by Bishnu and colleagues (2019). This study demonstrated the ability of metformin to hamper the development of resistance through increased drug sensitivity, decreased proliferation, and reduced migratory abilities of the resistant cells. Metformin treatment also decreased the cancer stem cell population, expression of specific biomarkers and pluripotent genes (Bishnu et al. 2019).

Importantly, another anti-inflammatory property of metformin is the ability to partly reversed cisplatin-stimulated IL-6 secretion in the stromal fibroblasts of patients via suppressing NFκB signaling. In addition, attenuated fibroblast-facilitated tumor growth in 3D organotypic cocultures and murine xenograft models (Xu et al. 2018). It has been established that tumor microenvironment was proven to contribute tremendously to the unsatisfactory chemotherapeutic efficacy in ovarian cancer. Consistent with this, metformin increased the sensitivity of metastatic cells, isolated from a female patient with extensive invasive ovarian cancer, to platinum-based chemotherapy, and this effect was dose-dependent (Y. Liu et al. 2018).

Studies using various cancer models, including breast and ovarian cancer have demonstrated the efficacy of metformin in attenuating cancer stem cells (CSCs) through the targeting of specific pathways involved in cell differentiation, proliferation, invasion and metabolism. CSCs are associated with relapse, metastasis, and drug resistance (Alvero et al. 2009;

Luo et al. 2015; Lupia and Cavallaro 2017). Shank et al. (2012) demonstrated that metformin inhibited the formation of CSC tumor spheres from both cell lines and patient tumors. In addition, metformin significantly increased the ability of cisplatin to restrict tumor *in vivo* (Shank et al. 2012). A later study also observed that in ovarian cancer cell lines low doses of metformin selectively inhibit CSCs and EMT. Additionally, metformin decreased ovarian tumor xenograft growth and enhanced effect of cisplatin (Zhang et al. 2015).

Rocha and colleagues (2011) found that combination regimen with metformin and paclitaxel synergistically inhibited the proliferation of luminal A MCF-7 cells compared with individual drug treatments and that the AMPK signaling pathway was involved in this process (Rocha et al. 2011). Furthermore, studies carried out by Liu and colleagues showed that metformin and mTOR inhibitor everolimus sensitize breast cancer cells to cytotoxic effects of chemotherapeutic agents, such as carboplatin, paclitaxel and doxorubicin, *in vitro* (Liu et al. 2012b). Metformin combined with everolimus had a greater effect of apoptosis and proliferation inhibition on breast cancer than monotherapy (Wang et al. 2015) and the same effect was also found in the combination treatment of metformin with rapamycin (Zakikhani et al. 2010). In addition, our group reported that under *in vitro* conditions 10 μM metformin along with doxorubicin and paclitaxel synergistically inhibit proliferation in TNBC cells (Guimaraes et al. 2015). Another study observed that, in mouse xenograft models of breast cancer cell lines, metformin had a synergistic effect when combined with paclitaxel, carboplatin or doxorubicin (Iliopoulos, Hirsch, and Struhl 2011).

Metformin inhibition of CSCs was first demonstrated in 2009 *in vitro* and *in vivo* breast cancer models. Hirsch et al. (2009) showed that low doses metformin (100-300 μM) preferentially kill CSCs ($CD44^{high}$/ $CD24^{low}$) in different subtypes of breast cancer whereas these doses failed to affect non-stem cancer cells. Indeed, in a xenograft mouse model metformin combined with doxorubicin reduces tumor mass and prevents recurrence more effectively when compared with single-agent therapy (Hirsch et al. 2009). Similarly, other studies have also demonstrated that metformin may kill breast CSCs and reduce the expressions of CSC markers, such as, those of $CD44^{high}/CD24^{low}$, ALDH-1, or OCT4 (Jung et al. 2011; Song et al. 2012). Other emerging data also indicate that metformin significantly decreased the percentage of CSC in triple negative cells HCC1806 and HCC1937 (Shi et al. 2017).

In addition to effects in monotherapy treatment, it was found that metformin combination therapy with the anti-HER2 monoclonal antibody trastuzumab could reduce CSC in *HER2*-amplified breast cancer cells (Vazquez-Martin et al. 2011). Another study demonstrated the favorable effects of metformin treatment in breast cancer to overcome resistance of trastuzumab. Metformin induced preferential killing of CSC subpopulations and the combination of trastuzumab and metformin induced a decrease of the xenograft's tumor size which was greater than either trastuzumab or metformin alone (Cufi et al. 2012).

Metformin synergized with 5-fluorouracil, epirubicin, and cyclophosphamide (FEC) in non-stem breast cancer cells and breast CSC. Interestingly, these effects occur through an AMPK-dependent mechanism in parental breast cancer and AMPK-independent manner in breast CSC (Soo et al. 2015). In 2014, Qu and colleagues showed the reversal of multidrug resistance in breast cancer cells through activation of AMPK/mTOR by metformin. The authors also reported that metformin reversed EMT phenotype and decreased the invasive capacity of chemoresistance MCF7/5-FU cells and in TNBC cells (Qu et al. 2014).

Chemosensitizing effects of metformin also been confirmed in a study that investigated the effects of metformin treatment in the prevention of a doxorubicin-chemoresistant phenotype in luminal A and TNBC cell lines. Metformin pretreatment generates more oxidative stress and significantly prevented the establishment of the doxorubicin-resistant phenotype in both luminal A and TNBC cell lines (Marinello et al. 2019).

In addition, metformin has additively effects with tamoxifen on reducing tumor growth *in vitro* and *in vivo* in ER-positive breast cancer cells through the Bax/Bcl-2 apoptotic pathway and the AMPK/mTOR growth pathway (Ma et al. 2014). Also, metformin inhibited tumor cell migration and invasion in tamoxifen-resistant breast cancer cells (Jang et al. 2014). Another study confirmed that metformin may be one of the effective therapeutic agents on tamoxifen-resistant breast cancer by repression of expression and transcriptional activity of ERα (Kim et al. 2016).

Moreover, in the tumor microenvironment, it has been shown that metformin enhances the number and function of CD8+ tumor-infiltrating lymphocytes leading to a cytotoxic response against cancer cells (Eikawa et al. 2015). Besides its effect on T cells, metformin is able promote macrophage reprogramming from an M2 to M1-like phenotype, an antitumor phenotype, through the AMPK pathway (Ding et al. 2015). Interestingly, tumor tissues from patients with TNBC were found to have an increased number of tumor-infiltrating lymphocytes accompanied by an increased programmed cell death ligand (PD-L1) levels (Wimberly et al. 2015). Blocking the interaction between the PD-1 protein and one of its ligands, PD-L1, has been reported to have impressive antitumor responses. Recently, Cha et al. (2018) reported that metformin mediated the blocking of PD-L1/PD-1 signaling enhancing the cytotoxic T lymphocyte activity against cancer cells. AMPK activated by metformin decreases PD-L1 levels in cancer cells by altering PD-L1 glycan structure, which promotes PD-L1 degradation and subsequently blocking immune inhibitory signaling. Consistently, metformin-treated patients have reduced PD-L1 levels with increased AMPK activation. Also, metformin increased the killing effects of cytotoxic T cells in two TNBC cell lines, MDA-MB-231 and BT-549 (Cha et al. 2018).

Reduction of Cancer Risk by Metformin: Observational Studies

Diabetes and cancer are both chronic diseases of high prevalence and mortality. Several studies indicate that patients with T2DM also have a higher propensity to acquire certain types of cancer (Zendehdel 2003; Saltzman et al. 2007; Soranna et al. 2012; Badrick and Renehan 2014). As mentioned before, this association may be doubtful because you cannot be sure if it is due to diabetes or other prognostic factors. Diabetic patients were associated with an increased risk of ovarian cancer, compared with no diabetic subjects (Lee et al. 2013). Moreover, diabetic women are more likely to have poorly differentiated tumor and poorer clinical outcome when compared with non-diabetic women (Bakhru, Buckanovich, and Griggs 2011). In addition, diabetes is associated with increased breast cancer incidence and mortality (De Bruijn et al. 2013; Boyle et al. 2012; Pamela J. Goodwin et al. 2002). The meta-analysis of forty studies has founded a significantly increased risk of breast cancer associated with diabetes in women (Hardefeldt, Edirimanne, and Eslick 2012). Later, another meta-analysis showed a moderately increased risk of BC in diabetic patients (Bota, Autier, and Boyle 2018). Nevertheless, a recent study published on *Journal of Diabetes Investigation* has shown that postmenopausal women with breast cancer are more likely to develop diabetes (C. Wang, Shih, and Huang 2019).

In 2005 Evans and colleagues first reported that the use of metformin in T2DM patients was associated with a lower overall incidence of cancer compared with patients treated with other anti-diabetic drugs (Evans et al. 2005). Epidemiologic studies suggest that metformin use may be associated not only with reduced cancer incidence, but also mortality. A prospective cohort study conducted in 2010, evaluated the association between the use of metformin and cancer mortality in 1,353 patients with T2DM, and it confirmed that the cancer mortality risk for metformin users is lower compared with that of nonusers (Landman et al. 2010). Moreover, this direct relation between decrease of cancer mortality and exposure to metformin was in a dose dependent manner (Landman et al. 2010). More recently, a meta-analysis showed that patients with cancer and T2DM had beneficial effects in overall survival and cancer-specific survival than patients treated with other glucose-lowering drugs (Yin et al. 2013).

Diabetic patients receiving medications other than metformin had a slightly higher incidence of breast compared to those who received metformin. The association was observed for cancers positive for both estrogen receptor and progesterone receptor and those that were negative for HER2 (Chlebowski et al. 2012). A protective effect of metformin against breast cancer in postmenopausal diabetic women was indicated a meta-analysis of breast cancer with diabetes (Col et al. 2012). Moreover, epidemiological studies on T2DM patients demonstrated an association between the use of metformin and reduced breast cancer incidence and mortality with respect to other antidiabetic drugs, especially in overweight/insulin-resistant women (Gandini et al. 2014).

In regarding of subtypes of breast cancer, metformin use in TNBC patients was not associated with beneficial effects on distant metastasis-free survival, recurrence-free survival or overall survival. Nevertheless, patients treated with metformin had a slightly lower risk of distant metastases when compared with diabetics not receiving metformin and non-diabetics (Bayraktar et al. 2012).

Analysis of the Adjuvant Lapatinib and/or Trastuzumab Treatment Optimization (ALTTO) phase III randomized trial showed that metformin exerted a statistically-significant beneficial effect in disease-free survival and overall survival in those patients with diabetes who had primary HER2-positive and hormone receptor-positive breast cancer. Also, results from this trial demonstrated that women with breast cancer and diabetes previous treated with insulin were worse outcomes than those of women treated with metformin (Sonnenblick et al. 2017).

Some epidemiological studies and meta-analysis have shown that metformin use is associated with decreased risk and/or reduced mortality of ovarian cancer. Epidemiological case–control study carried out by Bodmer et al. (2011) showed that long-term use of metformin was associated with lower rates of ovarian cancer (Bodmer et al. 2010). A further case–control analysis found an association between metformin and improved survival from epithelial ovarian cancer (Kumar et al. 2013). A retrospective cohort study reported that women with diabetes who used metformin had a decreased risk of ovarian cancer compared with those who did not use it (Tseng 2015). Furthermore, meta-analysis study confirmed that metformin decrease the occurrence of ovarian cancer among diabetic patients (Dilokthornsakul et al. 2013). Zhang et al. also showed that metformin was associated with lower risk of all-cause mortality in ovarian cancer (Z. J. Zhang and Li 2014).

Effects of Metformin Breast and Ovarian Cancer: Clinical Studies and Trials

Encouraged by positive results in these preclinical studies, there are 50 studies registered on Clinical Trials (www.clinicaltrials.org, accessed 01 July 2019) using the search query “metformin” and “breast cancer” or “ovarian cancer”. These trials evaluate metformin in monotherapy or in combination with standard chemotherapy to treat patients in different cancer stages. From 2011 up to date, only eight clinical trials were registered studying metformin related to ovarian cancer, of which only two completed phase I and phase II trials. The low number of trials regarding ovarian cancer can be partially understood and explained by the high mortality rate of the disease (Momenimovahed et al. 2019). On the other hand, from 2008 to 2019, 43 studies have been placed to examine the effect of metformin on patients related to breast cancer. To date, six phase I and six phase II trials have the status of “completed” studies. These numbers confirm the amount of effort to investigate metformin as a potential anti-cancer agent.

A neoadjuvant, single arm, “window of opportunity” trial (NCT00897884) hypothesized that metformin could reduce cell proliferation rates in tumor tissue. The method consisted to take women with untreated breast cancer and give 500 mg of metformin three times daily for more than 2 weeks after diagnostic biopsy until surgery. The results showed that levels of insulin receptor ($P = 0.04$), phosphorylation status of Akt (S473) ($P = 0.0001$), ERK1/2 (T202/Y204) ($P < 0.0001$), AMPK (T172) ($P < 0.005$), and Acetyl-coA (S79) ($P = 0.02$) decreased significantly in tumor. Part of the authors conclusion is the suggestion of insulin-dependent effects being important in the clinical settings (Dowling et al. 2015).

Table 2. Observational studies investigating the association between metformin risk of breast and ovarian cancer

Type of cancer	Population	Outcome measures	Conclusions	Reference
Breast cancer	4,085 metformin users	Evaluate the association of metformin therapy with cancer incidence	results were not all statistically significant.	(Libby et al. 2009)
Breast cancer	315,890 incident diabetic individuals	Evaluate the association of metformin therapy with cancer incidence	No significant association of metformin treatment with the incidence of breast cancer	(Dankner et al. 2019)
Early-stage breast cancer	2,529 females 68 diabetic patients taking metformin, 87 diabetic patients not taking metformin, and 2,374 non- diabetic patients	Evaluate the influence of metformin therapy on breast cancer prognosis	Metformin improve pathologic complete response rates but it did not significantly improve survival	(Jiralerspong et al. 2009)
Breast cancer	22,621diabetics females using oral antidiabetic drugs	time-response analysis, of hypoglycemic drugs in risk of breast cancer	Long-term use (>5 years) of metformin decreases breast cancer risk	(Bodmer et al. 2010)
Breast cancer	62,809	Tumor progression	Metformin did not affect the risk of breast; insulin-based regimens had the highest risk of progression to solid tumor	(C. J. Currie, Poole, and Gale 2009)
Breast cancer	112,408 individuals,	Impact of metformin on survival	metformin monotherapy reduced breast cancer mortality	(Craig J. Currie et al. 2013)

Type of cancer	Population	Outcome measures	Conclusions	Reference
Breast and ovarian cancer	145,826 women	Examine associations with diabetes diagnosis or duration of medication use.	Postmenopausal diabetics patients had a higher risk of reproductive tract and breast cancer and those who use metformin may be at lower risk; risk of death from breast cancer was lower for metformin users; decreased risk appears to be associated with length of use	(Gong et al. 2016)
Ovarian cancer	479,475 patients metformin ever-users (n = 193369) and never-users (n = 286106)	Whether ovarian cancer risk is modulate by metformin therapy in diabetics Asian patients	Metformin reduces risk of ovarian cancer in female patients with T2DM.	(Tseng 2015)
Ovarian cancer	571 patients	effects of metformin on longer progression-free survival and overall survival between ovarian cancer patients with type 2 diabetes	Greater survival rates was observed for ovarian cancer patients with type 2 diabetes using metformin	(S.-B. Wang et al. 2017)
Ovarian cancer	137 643 women over 40 years old and diagnosed with T2	Incidence of ovarian cancer in metformin users	No significant association between the use of metformin and the incidence of ovarian cancer in women with T2D was found.	(Urpilainen et al. 2018)
Ovarian cancer	421 women with T2D	Prognosis of ovarian cancer in women with type 2 diabetes treated with metformin or other forms of antidiabetic drug	Inconclusive results regarding association between metformin and ovarian cancer survival	(Urpilainen et al. 2018)
Ovarian cancer	2291 patients	Metformin effects on survival in patients with ovarian, fallopian tube or primary peritoneal cancer.	No statistically significant association was observed between metformin use and overall survival[i]	(Garcia et al. 2017)
Ovarian cancer	143 ovarian cancer patients	Metformin on the risk of ovarian cancer recurrence and mortality.	Metformin prolonged recurrence free survival	(Bar et al. 2016)

A phase I study involving breast cancer, posted in March 2008 and completed in 2012 (NCT00659568), was based on the idea that metformin and temsirolimus could stop the growth of cancer cells by blocking some of the enzymes needed for cell growth. Thus, the aim was to establish the maximum tolerated dose and define dose-limiting toxicities of this combination. Eleven patients were enrolled. The maximum tolerated dose (level -1) was 20 mg temsirolimus weekly and 500 mg daily of metformin (MacKenzie et al. 2012). Also, on phase I, another trial (NCT02145559) published results testing the hypothesis of metformin potentiating mTOR inhibition by sirolimus. The 24 patients enrolled were selected on day 7 to either sirolimus with metformin or sirolimus until day 21. Primary, it was compared the change in phsopho-p70S6K in peripheral blood mononuclear cells, which showed no significant difference (- 0.12 vs. – 0.16; P = 0.64), as well to other serum pharmacodynamic biomarkers. They conclude that, although well tolerated, the association did not improve the therapy (Sehdev et al. 2018).

There are three ongoing phase I trials. One of the studies which phase I was completed in May 2019 is investigating the possible effects on women with a history of early stage breast cancer after receiving a combination of metformin and omega-3 fatty acids up to one year (NCT02278965). The other two are recruiting patients newly diagnosed to be treat with metformin and atorvastatin to determine the effects on proliferation during the interval between breast biopsy and surgery (NCT01980823) and patients to evaluate the safety, tolerability, and pharmacokinetics of a new drug, named GDC-0077, in combination with fulvestrant, palbociclib and metformin (NCT03006172).

Other aspects regarding the use of metformin has been tested on clinical trial phase I, but not yet published, for example: i) to measure if metformin is taken up into the breast tumor cells (NCT02882581); ii) to establish the maximum tolerated combined dosing of erlotinib and metformin as well as deciding if there is a potential clinical utility (NCT01650506); iii) to determine the changes in the signal pathway profiling of breast tissue of overweight or obese women at elevated risk for breast cancer (NCT01793948); iv) to find the highest tolerable dose of metformin plus rosiglitazone that can be given in combination with exemestane to patients who are obese and postmenopausal with hormone-receptive-positive breast cancer that has spread to other parts of the body (NCT00933309).

Phase II clinical trials is focused on the therapeutic efficacy into patient population to establish whether the drug may ultimately benefit patients and should be studied in a phase III trial. There are at least eight studies published relating breast cancer and metformin on stage II trial. In 2014, two studies were first published.

The first article is a prospective, multicenter, phase II randomized, placebo-controlled trial intended to identify the anti-tumor effect of metformin with preoperatively given letrozole (Experimental arm: Metformin HCl 500 mg (morning/evening) PO daily x 1week Metformin HCl 1000 mg (morning) and 500 mg (evening) PO daily x 1week Metformin HCl 1000 mg (morning/evening) PO daily x 22 weeks), in 208 postmenopausal estrogen receptor positive breast cancer patients (NCT01589367). The clinical response rate, comparing with RECIST 1.1 from baseline to 24 week after completing medication will be assessed along with other factors, such as ki-67changes (Jisun Kim et al. 2014). The second trial article (NCT00930579) relied on a pre surgical intervention model to evaluate the tissue-related effects of metformin in overweight/ obese breast cancer patients. Patients received metformin 1,500 mg per day for at least 2 weeks and was to be taken until the night prior to surgery. The study had 33 patients completing the trial and considered evaluable. The authors conclude that there was no significant reduction in tumor proliferation, nevertheless they added that host factors, such as baseline ki-67, should be considered in the next studies, given the differences between patients treated with metformin (Kalinsky et al. 2014).

The increment of clinical phase II studies along with years has helped to accumulate clinical experience and significantly shed light into clinical evaluation path of metformin in cancer prevention and treatment. Concisely mentioning the evolution through the years, it is been published studies testing i) the efficacy and safety of adjuvant metformin for operable breast cancer patients with overweight or pre-diabetes mellitus

(NCT00909506), concluding that the results indicate an anticancer potential; ii) the potential effects on reduction of obesity-associated breast cancer risk (NCT02028221); iii) the hypothesis that metformin could further increase the efficacy of aromatase inhibitors in patients with metastatic breast cancer after the failure of the first line endocrine therapy (NCT01654185), where it failed to prove an improved efficacy, although an excellent tolerability was verified; iv) the pharmacodynamic effects of metformin on breast cancer metabolism, using PET-CT scan and blood testes carried out before and after the 2 week period treatment (NCT01266486), which found two types of breast cancer metabolic response to metformin; and v) the efficacy of metformin conjugated with doxorubicin and cyclophosphamide, having as primary endpoint progression-free survival (NCT01885013), unfortunately no evidences were found to support the combination as first line chemotherapy in metastatic breast cancer (Ko et al. 2015; Martinez et al. 2016; Y. Zhao et al. 2017; Lord et al. 2018; Nanni et al. 2019). Although all results presented so far did not change abruptly the chemotherapy regimens, it unravels new paths and guides next studies, which includes a controversial test in non-diabetic cancer patients.

It's based on previously studies that metformin has reached phase III on clinical trials. Together, these annotations motived the creation of NCIC Clinical Trials Group (NCIC CTG) MA.32, a multicenter phase III study (NCT01101438) where patients are stratified according to hormone-receptor status, body mass index, HER2 status, and prior chemotherapy and receive oral metformin hydrochloride twice daily (once daily in weeks 1-4) for up to 5 years in the absence of disease progression or unacceptable toxicity. The primary outcome measured is invasive disease-free survival in hormone receptor negative and positive sub-groups, but also conducted analysis on weight and metabolic factors at six months in the first 492 patients with paired blood samples. It was possible to observe that at six months, weight and blood variables had a statistically significantly decrease in the metformin treated group. This confirms previous trials that metformin affects body weight and several metabolic factors associated with insulin resistance and, more importantly, provided clear evidence that metformin improves the metabolic profile of non-diabetic breast cancer patients (P. J. Goodwin et al. 2015).

There is yet one ongoing phase III clinical trial, with recruiting status. This aims to elucidate the use of metformin in preventing breast cancer in patients with atypical hyperplasia or *in situ* breast cancer (NCT01905046). Interestingly, two articles published explored the data generated from clinical trial phase III, NCT00490139 and NCT01101438. Briefly, their conclusion are that in diabetic patients with hormone receptor-positive and HER2-positive breast cancer, metformin had a salutary effect (Sonnenblick et al. 2017) and after six months of treatment with metformin, non-diabetic patients with high risk for breast cancer showed a biochemical vitamin B12 deficiency; not related with anemia (Lohmann et al. 2017), respectively.

It is interesting that ovarian cancer is the most lethal gynecological cancer (Momenimovahed et al. 2019), however limited data is available in comparison with breast cancer. As mentioned before, there are only eight clinical trials registered on clinical trials. Only two studies have the "completed" status, one phase I and one phase II; five trials are posted as "recruiting" status, one phase I, three phase II, and one phase III; one phase I trial has the status of withdrawn. It will be presented a brief summary of what these studies are intent to and, when available, the results obtained.

A trial named as IMPACT, an early phase I (NCT03378297), is conducted in women with advanced ovarian cancer going through a diagnostic laparoscopy. A treatment is given for 10-14 days before surgery, metformin tablets – 850 mg x 2 orally. The goal is to detect changes in the expression of specific biomarkers. Also, on phase I, NCT02312661 is a single-center, dose-escalation trial with a traditional escalation rule to determine the recommended phase II dose of metformin in combination with carboplatin/paclitaxel in ovarian cancer patients. The authors already published their results and conclude that the recommended phase II dose of metformin is 1000 mg (Broekman et al. 2018).

Recently, the first prospective study of metformin in epithelial ovarian cancer patients phase II trial (NCT01579812) was published. Thirty-two patients were divided into two groups: treated with neoadjuvant metformin followed primary debulking surgery and adjuvant metformin plus chemotherapy and neo-adjuvant chemotherapy plus chemotherapy, followed by interval debulking and adjuvant chemotherapy plus metformin. The parameters evaluated were side effects, progression free survival and overall survival, as well the presence of cancer stem-like cells for metformin treated tumors. Solid results were obtained, more importantly they found a 3-fold decrease in aldehyde dehydrogenase and cancer stem-like cells at baseline. This study corroborated with translational studies showing impact of metformin on cancer stem-like cells and paved the way of metformin to reach phase III studies (Buckanovich et al. 2017).

Ongoing phase II trials, there is an open-label, non-randomized, pilot study of metformin combined with paclitaxel and carboplatin assessing the safety, toxicity, and progression free survival of advanced stage ovarian carcinoma patients expected to be completed in 2021 (NCT02437812). With an estimated completion date around 2022, the trial NCT02122185 studies if the increment of metformin to standard adjuvant or neoadjuvant chemotherapy and extended regimen of metformin after chemotherapy improves progression free survival in non-diabetic subjects with stage III or stage IV ovarian, primary peritoneal, or fallopian tube carcinoma. And a new approach to the treatment of early stage endometrial cancer has been tested and is likely to review the results in 2020. The idea is combining three interventions, estrogen replacement therapy plus metformin plus weight loss. The outcome expected is pathological complete response in a time frame of 6 months (NCT01686126).

Lastly, a phase III trial ongoing since 2017 is a global cancer study to determine effectiveness of different treatments (NCT02201381), including oral metformin up to 1000 mg, evaluating the overall survival (years and months) up to 5 years. The proposal of the study is to be a real world setting and develop an exploratory analysis on the relationship between the degree of response and changes in biochemical markers. This study is estimated to be completed by October 22, 2023.

Conclusion and Perspectives

In recent years there has been growing interest in drug repurposing of current medications especially in treatment of poor prognosis diseases, such as triple-negative breast cancer and high grade serous ovarian cancer. The evidence collected so far clearly shows that, metformin a well-tolerated, inexpensive medication with a long history of clinical experience in treating diabetes, appears as an effective antineoplastic drug. Despite extensive research about the molecular effects of metformin in cancer cells, the precise molecular target and mechanism of action remains controversial. A deeper knowledge about these mechanisms could improve the potential use of metformin in cancer and potential clinical markers could be better identified. In addition, although the combination of metformin with anticancer agents shows a synergistic efficiency in breast and ovarian cancers, further investigations, including placebo controlled and randomized clinical trials are needed to confirm its anti-neoplastic properties.

References

Addington, James, and Miriam Freimer. 2016. "Chemotherapy-Induced Peripheral Neuropathy: An Update on the Current Understanding." *F1000Research* 5 (June): 1466. https://doi.org/10.12688/f1000research.8053.1.

Alimova, Irina N., Bolin Liu, Zeying Fan, Susan M. Edgerton, Thomas Dillon, Stuart E. Lind, and Ann D. Thor. 2009. "Metformin Inhibits Breast Cancer Cell Growth, Colony Formation and Induces Cell Cycle Arrest *in Vitro*." *Cell Cycle* 8 (6): 909–15. https://doi.org/10.4161/cc.8.6.7933.

Alshawi, Ahmed, and Loranne Agius. 2019. "Low Metformin Causes a More Oxidized Mitochondrial NADH/NAD Redox State in Hepatocytes and Inhibits Gluconeogenesis by a Redox-Independent

Mechanism." *Journal of Biological Chemistry* 294 (8): 2839–53. https://doi.org/10.1074/jbc.RA118.006670.

Alvero, Ayesha B., Rui Chen, Han-Hsuan Fu, Michele Montagna, Peter E. Schwartz, Thomas Rutherford, Dan-Arin Silasi, et al. 2009. "Molecular Phenotyping of Human Ovarian Cancer Stem Cells Unravels the Mechanisms for Repair and Chemoresistance." *Cell Cycle* 8 (1): 158–66. https://doi.org/10.4161/cc.8.1.7533.

Anisimov, Vladimir N., Lev M. Berstein, Peter A. Egormin, Tatiana S. Piskunova, Irina G. Popovich, Mark A. Zabezhinski, Irina G. Kovalenko, et al. 2005. "Effect of Metformin on Life Span and on the Development of Spontaneous Mammary Tumors in HER-2/Neu Transgenic Mice." *Experimental Gerontology* 40 (8–9): 685–93. https://doi.org/10.1016/j.exger.2005.07.007.

Annenkov, Alexander. 2009. "The Insulin-Like Growth Factor (IGF) Receptor Type 1 (IGF1R) as an Essential Component of the Signalling Network Regulating Neurogenesis." *Molecular Neurobiology* 40 (3): 195–215. https://doi.org/10.1007/s12035-009-8081-0.

Ariaans, Gerke, Mathilde Jalving, Emma Geertruida Elisabeth de Vries, and Steven de Jong. 2017. "Anti-Tumor Effects of Everolimus and Metformin Are Complementary and Glucose-Dependent in Breast Cancer Cells." *BMC Cancer* 17 (1): 232. https://doi.org/10.1186/s12885-017-3230-8.

Aroda, Vanita R., Sharon L. Edelstein, Ronald B. Goldberg, William C. Knowler, Santica M. Marcovina, Trevor J. Orchard, George A. Bray, et al. 2016. "Long-Term Metformin Use and Vitamin B12 Deficiency in the Diabetes Prevention Program Outcomes Study." *The Journal of Clinical Endocrinology & Metabolism* 101 (4): 1754–61. https://doi.org/10.1210/jc.2015-3754.

Badrick, Ellena, and Andrew G. Renehan. 2014. "Diabetes and Cancer: 5years into the Recent Controversy." *European Journal of Cancer* 50 (12): 2119–25. https://doi.org/10.1016/j.ejca.2014.04.032.

Bailey, C. J., and C. Day. 1989. "Traditional Plant Medicines as Treatments for Diabetes." *Diabetes Care* 12 (8): 553–64. https://doi.org/10.2337/diacare.12.8.553.

Bailey, C. J., C. Wilcock, and J. H. B. Scarpello. 2008. "Metformin and the Intestine." *Diabetologia* 51 (8): 1552–53. https://doi.org/10.1007/s00125-008-1053-5.

Bailey, Clifford J. 2017. "Metformin: Historical Overview." *Diabetologia* 60 (9): 1566–76. https://doi.org/10.1007/s00125-017-4318-z.

Bakhru, Arvind, Ronald J. Buckanovich, and Jennifer J. Griggs. 2011. "The Impact of Diabetes on Survival in Women with Ovarian Cancer." *Gynecologic Oncology* 121 (1): 106–11. https://doi.org/10.1016/j.ygyno.2010.12.329.

Bar, Daniel, Ofer Lavie, Nili Stein, Ido Feferkorn, and Ayelet Shai. 2016. "The Effect of Metabolic Comorbidities and Commonly Used Drugs on the Prognosis of Patients with Ovarian Cancer." *European Journal of Obstetrics & Gynecology and Reproductive Biology* 207 (December): 227–31. https://doi.org/10.1016/j.ejogrb.2016.09.005.

Barnett, Anthony H. 2010. "Avoiding Hypoglycaemia While Achieving Good Glycaemic Control in Type 2 Diabetes through Optimal Use of Oral Agent Therapy." *Current Medical Research and Opinion* 26 (6): 1333–42. https://doi.org/10.1185/03007991003738063.

Baur, Joseph A., and Morris J. Birnbaum. 2014. "Control of Gluconeogenesis by Metformin: Does Redox Trump Energy Charge?" *Cell Metabolism* 20 (2): 197–99. https://doi.org/10.1016/j.cmet.2014.07.013.

Bayraktar, Soley, Leonel F. Hernadez-Aya, Xiudong Lei, Funda Meric-Bernstam, Jennifer K. Litton, Limin Hsu, Gabriel N. Hortobagyi, and Ana M. Gonzalez-Angulo. 2012. "Effect of Metformin on Survival Outcomes in Diabetic Patients with Triple Receptor-Negative Breast Cancer." *Cancer* 118 (5): 1202–11. https://doi.org/10.1002/cncr.26439.

Bishnu, Aniketh, Asmita Sakpal, Nilanjana Ghosh, Priyanka Choudhury, Koel Chaudhury, and Pritha Ray. 2019. "Long Term Treatment of Metformin Impedes Development of Chemoresistance by Regulating Cancer Stem Cell Differentiation through Taurine Generation in Ovarian Cancer Cells." *The International Journal of Biochemistry & Cell Biology* 107 (February): 116–27. https://doi.org/10.1016/j.biocel.2018.12.016.

Bodmer, M., C. Meier, S. Krahenbuhl, S. S. Jick, and C. R. Meier. 2010. "Long-Term Metformin Use Is Associated With Decreased Risk of Breast Cancer." *Diabetes Care* 33 (6): 1304–8. https://doi.org/10.2337/dc09-1791.

Bois, A. du. 2005. "2004 Consensus Statements on the Management of Ovarian Cancer: Final Document of the 3rd International Gynecologic Cancer Intergroup Ovarian Cancer Consensus Conference (GCIG OCCC 2004)." *Annals of Oncology* 16 (suppl_8): viii7–12. https://doi.org/10.1093/annonc/mdi961.

Bota, Maria, Philippe Autier and Peter Boyle. 2018. "The Risk of Breast Cancer in Women with Diabetes." *Diabetes* 67 (Supplement 1): 180-OR. https://doi.org/10.2337/db18-180-OR.

Boyle, P, M Boniol, A Koechlin, C Robertson, F Valentini, K Coppens, L-L Fairley, et al. 2012. "Diabetes and Breast Cancer Risk: A Meta-Analysis." *British Journal of Cancer* 107 (9): 1608–17. https://doi.org/10.1038/bjc.2012.414.

Bridges, Hannah R., Andrew J. Y. Jones, Michael N. Pollak, and Judy Hirst. 2014. "Effects of Metformin and Other Biguanides on Oxidative Phosphorylation in Mitochondria." *Biochemical Journal* 462 (3): 475–87. https://doi.org/10.1042/BJ20140620.

Broekman, K E, A K L Reyners, D J Touw, M A J Hof, and M Jalving. 2018. "Phase Ib Study of Metformin in Combination with Carboplatin/Paclitaxel Chemotherapy in Patients with Advanced Epithelial Ovarian Cancer." *Annals of Oncology* 29 (suppl_8): viii332–58. https://doi.org/10.1093/annonc/mdy285.199.

Bruijn, K. M. J. De, L. R. Arends, B. E. Hansen, S. Leeflang, R. Ruiter, and C. H. J. van Eijck. 2013. "Systematic Review and Meta-Analysis of the Association between Diabetes Mellitus and Incidence and Mortality in Breast and Colorectal Cancer." *British Journal of Surgery* 100 (11): 1421–29. https://doi.org/10.1002/bjs.9229.

Buckanovich, Ronald J., Jason Brown, Jessica Shank, Kent A. Griffith, R. Kevin Reynolds, Carolyn Johnston, Karen McLean, et al. 2017. "A Phase II Clinical Trial of Metformin as a Cancer Stem Cell Targeting Agent in Stage IIc/III/IV Ovarian, Fallopian Tube, and Primary

Peritoneal Cancer." *Journal of Clinical Oncology* 35 (15_suppl): 5556–5556. https://doi.org/10.1200/JCO.2017.35.15_suppl.5556.

Buse, John B., Ralph A. DeFronzo, Julio Rosenstock, Terri Kim, Colleen Burns, Sharon Skare, Alain Baron, and Mark Fineman. 2015. "The Primary Glucose-Lowering Effect of Metformin Resides in the Gut, Not the Circulation. Results From Short-Term Pharmacokinetic and 12-Week Dose-Ranging Studies." *Diabetes Care*, August, dc150488. https://doi.org/10.2337/dc15-0488.

Carter, D, H C S Howlett, N F Wiernsperger, and C J Bailey. 2003. "Differential Effects of Metformin on Bile Salt Absorption from the Jejunum and Ileum." *Diabetes, Obesity & Metabolism* 5 (2): 120–25. http://www.ncbi.nlm.nih.gov/pubmed/12630937.

Cha, Jong-Ho, Wen-Hao Yang, Weiya Xia, Yongkun Wei, Li-Chuan Chan, Seung-Oe Lim, Chia-Wei Li, et al. 2018. "Metformin Promotes Antitumor Immunity via Endoplasmic-Reticulum-Associated Degradation of PD-L1." *Molecular Cell* 71 (4): 606-620.e7. https://doi.org/10.1016/j.molcel.2018.07.030.

Chang, Suk-Joon, Melissa Hodeib, Jenny Chang, and Robert E. Bristow. 2013. "Survival Impact of Complete Cytoreduction to No Gross Residual Disease for Advanced-Stage Ovarian Cancer: A Meta-Analysis." *Gynecologic Oncology* 130 (3): 493–98. https://doi.org/10.1016/j.ygyno.2013.05.040.

Chitnis, M. M., J. S. P. Yuen, A. S. Protheroe, M. Pollak, and V. M. Macaulay. 2008. "The Type 1 Insulin-Like Growth Factor Receptor Pathway." *Clinical Cancer Research* 14 (20): 6364–70. https://doi.org/10.1158/1078-0432.CCR-07-4879.

Chlebowski, Rowan T., Anne McTiernan, Jean Wactawski-Wende, JoAnn E. Manson, Aaron K. Aragaki, Thomas Rohan, Eli Ipp, et al. 2012. "Diabetes, Metformin, and Breast Cancer in Postmenopausal Women." *Journal of Clinical Oncology* 30 (23): 2844–52. https://doi.org/10.1200/JCO.2011.39.7505.

Christensen, Mette M.H., Charlotte Brasch-Andersen, Henrik Green, Flemming Nielsen, Per Damkier, Henning Beck-Nielsen, and Kim Brosen. 2011. "The Pharmacogenetics of Metformin and Its Impact on

Plasma Metformin Steady-State Levels and Glycosylated Hemoglobin A1c." *Pharmacogenetics and Genomics* 21 (12): 837–50. https://doi.org/10.1097/FPC.0b013e32834c0010.

Christopoulos, Panagiotis F, Pavlos Msaouel, and Michael Koutsilieris. 2015. "The Role of the Insulin-like Growth Factor-1 System in Breast Cancer." *Molecular Cancer* 14 (February): 43. https://doi.org/10.1186/s12943-015-0291-7.

Cignarelli, Angelo, Valentina Annamaria Genchi, Irene Caruso, Annalisa Natalicchio, Sebastio Perrini, Luigi Laviola, and Francesco Giorgino. 2018. "Diabetes and Cancer: Pathophysiological Fundamentals of a 'Dangerous Affair.'" *Diabetes Research and Clinical Practice* 143 (September): 378–88. https://doi.org/10.1016/j.diabres.2018.04.002.

Col, Nananda F, Leslie Ochs, Vicky Springmann, Aaron K Aragaki, and Rowan T Chlebowski. 2012. "Metformin and Breast Cancer Risk: A Meta-Analysis and Critical Literature Review." *Breast Cancer Research and Treatment* 135 (3): 639–46. https://doi.org/10.1007/s10549-012-2170-x.

Cubeddu, L X, H Bönisch, M Göthert, G Molderings, K Racké, G Ramadori, K J Miller, and H Schwörer. 2000. "Effects of Metformin on Intestinal 5-Hydroxytryptamine (5-HT) Release and on 5-HT3 Receptors." *Naunyn-Schmiedeberg's Archives of Pharmacology* 361 (1): 85–91. http://www.ncbi.nlm.nih.gov/pubmed/10651152.

Cufi, Silvia, Bruna Corominas-Faja, Alejandro Vazquez-Martin, Christina Oliveras-Ferraros, Joan Dorca, Joaquim Bosch-Barrera, Begoña Martin-Castillo, and Javier A Menendez. 2012. "Metformin-Induced Preferential Killing of Breast Cancer Initiating CD44+CD24-/Low Cells Is Sufficient to Overcome Primary Resistance to Trastuzumab in HER2+ Human Breast Cancer Xenografts." *Oncotarget* 3 (4): 395–98. https://doi.org/10.18632/oncotarget.488.

Currie, C. J., C. D. Poole, and E. A. M. Gale. 2009. "The Influence of Glucose-Lowering Therapies on Cancer Risk in Type 2 Diabetes." *Diabetologia* 52 (9): 1766–77. https://doi.org/10.1007/s00125-009-1440-6.

Currie, Craig J., Chris D. Poole, Marc Evans, John R. Peters, and Christopher Ll. Morgan. 2013. "Mortality and Other Important Diabetes-Related Outcomes With Insulin vs Other Antihyperglycemic Therapies in Type 2 Diabetes." *The Journal of Clinical Endocrinology & Metabolism* 98 (2): 668–77. https://doi.org/10.1210/jc.2012-3042.

Currie, Erin, Almut Schulze, Rudolf Zechner, Tobias C. Walther, and Robert V. Farese. 2013. "Cellular Fatty Acid Metabolism and Cancer." *Cell Metabolism* 18 (2): 153–61. https://doi.org/10.1016/j.cmet.2013.05.017.

Dang, Jian-Hong, Zhi-Jun Jin, Xiao-Jun Liu, Dian Hu, Jing Wang, Yan Luo, and Ling-Ling Li. 2017. "Metformin in Combination with Cisplatin Inhibits Cell Viability and Induces Apoptosis of Human Ovarian Cancer Cells by Inactivating ERK 1/2." *Oncology Letters*, October. https://doi.org/10.3892/ol.2017.7176.

Dankner, Rachel, Nirit Agay, Liraz Olmer, Havi Murad, Lital Keinan Boker, Ran D Balicer, and Laurence S Freedman. 2019. "Metformin Treatment and Cancer Risk: Cox Regression Analysis with Time-Dependent Covariates of 320,000 Individuals with Incident Diabetes Mellitus." *American Journal of Epidemiology*, July. https://doi.org/10.1093/aje/kwz157.

Davidson, Jaime, and Harry Howlett. 2004. "New Prolonged-Release Metformin Improves Gastrointestinal Tolerability." *The British Journal of Diabetes & Vascular Disease* 4 (4): 273–77. https://doi.org/10.1177/14746514040040041101.

Dawed, Adem Y., Kaixin Zhou, Nienke van Leeuwen, Anubha Mahajan, Neil Robertson, Robert Koivula, Petra J. M. Elders, et al. 2019. "Variation in the Plasma Membrane Monoamine Transporter (PMAT) (Encoded by SLC29A4) and Organic Cation Transporter 1 (OCT1) (Encoded by SLC22A1) and Gastrointestinal Intolerance to Metformin in Type 2 Diabetes: An IMI DIRECT Study." *Diabetes Care* 42 (6): 1027–33. https://doi.org/10.2337/dc18-2182.

DeFronzo, Ralph, G. Alexander Fleming, Kim Chen, and Thomas A. Bicsak. 2016. "Metformin-Associated Lactic Acidosis: Current

Perspectives on Causes and Risk." *Metabolism* 65 (2): 20–29. https://doi.org/10.1016/j.metabol.2015.10.014.

Deng, Xin-Sheng, Shuiliang Wang, Anlong Deng, Bolin Liu, Susan M. Edgerton, Stuart E. Lind, Reema Wahdan-Alaswad, and Ann D. Thor. 2012. "Metformin Targets Stat3 to Inhibit Cell Growth and Induce Apoptosis in Triple-Negative Breast Cancers." *Cell Cycle* 11 (2): 367–76. https://doi.org/10.4161/cc.11.2.18813.

Dilokthornsakul, Piyemeth, Nathorn Chaiyakunapruk, Wichai Termrungruanglert, Chayanin Pratoomsoot, Surasak Saokeaw, and Rosarin Sruamsiri. 2013. "The Effects of Metformin on Ovarian Cancer: A Systematic Review." *International Journal of Gynecologic Cancer* 23 (9): 1544–51. https://doi.org/10.1097/IGC.0b013e3182a80a21.

Ding, Ling, Guikai Liang, Zhangting Yao, Jieqiong Zhang, Ruiyang Liu, Huihui Chen, Yulu Zhou, Honghai Wu, Bo Yang, and Qiaojun He. 2015. "Metformin Prevents Cancer Metastasis by Inhibiting M2-like Polarization of Tumor Associated Macrophages." *Oncotarget* 6 (34). https://doi.org/10.18632/oncotarget.5541.

Dowling, Ryan JO, Saroj Niraula, Martin C Chang, Susan J Done, Marguerite Ennis, David R McCready, Wey L Leong, et al. 2015. "Changes in Insulin Receptor Signaling Underlie Neoadjuvant Metformin Administration in Breast Cancer: A Prospective Window of Opportunity Neoadjuvant Study." *Breast Cancer Research* 17 (1): 32. https://doi.org/10.1186/s13058-015-0540-0.

Dujic, Tanja, Kaixin Zhou, Louise A. Donnelly, Roger Tavendale, Colin N. A. Palmer, and Ewan R. Pearson. 2015. "Association of Organic Cation Transporter 1 With Intolerance to Metformin in Type 2 Diabetes: A GoDARTS Study." *Diabetes* 64 (5): 1786–93. https://doi.org/10.2337/db14-1388.

Dujic, Tanja, Kaixin Zhou, Roger Tavendale, Colin N. A. Palmer, and Ewan R. Pearson. 2016. "Effect of Serotonin Transporter 5-HTTLPR Polymorphism on Gastrointestinal Intolerance to Metformin: A GoDARTS Study." *Diabetes Care* 39 (11): 1896–1901. https://doi.org/10.2337/dc16-0706.

Dunn, Christopher J., and David H. Peters. 1995. "Metformin." *Drugs* 49 (5): 721–49. https://doi.org/10.2165/00003495-199549050-00007.

Eikawa, Shingo, Mikako Nishida, Shusaku Mizukami, Chihiro Yamazaki, Eiichi Nakayama, and Heiichiro Udono. 2015. "Immune-Mediated Antitumor Effect by Type 2 Diabetes Drug, Metformin." *Proceedings of the National Academy of Sciences* 112 (6): 1809–14. https://doi.org/10.1073/pnas.1417636112.

El-Mir, Mohamad-Yehia, Véronique Nogueira, Eric Fontaine, Nicole Avéret, Michel Rigoulet, and Xavier Leverve. 2000. "Dimethylbiguanide Inhibits Cell Respiration via an Indirect Effect Targeted on the Respiratory Chain Complex I." *Journal of Biological Chemistry* 275 (1): 223–28. https://doi.org/10.1074/jbc.275.1.223.

Erices, R., M. L. Bravo, P. Gonzalez, B. Oliva, D. Racordon, M. Garrido, C. Ibanez, et al. 2013. "Metformin, at Concentrations Corresponding to the Treatment of Diabetes, Potentiates the Cytotoxic Effects of Carboplatin in Cultures of Ovarian Cancer Cells." *Reproductive Sciences* 20 (12): 1433–46. https://doi.org/10.1177/1933719113488441.

Erices, Rafaela, Maria Loreto Bravo, Pamela Gonzalez, Bárbara Oliva, Dusan Racordon, Marcelo Garrido, Carolina Ibañez, et al. 2013. "Metformin, at Concentrations Corresponding to the Treatment of Diabetes, Potentiates the Cytotoxic Effects of Carboplatin in Cultures of Ovarian Cancer Cells." *Reproductive Sciences* 20 (12): 1433–46. https://doi.org/10.1177/1933719113488441.

Evans, Josie M M, Louise A Donnelly, Alistair M Emslie-Smith, Dario R Alessi, and Andrew D Morris. 2005. "Metformin and Reduced Risk of Cancer in Diabetic Patients." *BMJ* 330 (7503): 1304–5. https://doi.org/10.1136/bmj.38415.708634.F7.

Ferlay, J., M. Colombet, I. Soerjomataram, C. Mathers, D. M. Parkin, M. Piñeros, A. Znaor, and F. Bray. 2018. "Estimating the Global Cancer Incidence and Mortality in 2018: GLOBOCAN Sources and Methods." *International Journal of Cancer*, December, ijc.31937. https://doi.org/10.1002/ijc.31937.

Finneran, Matthew M., and Mark B. Landon. 2018. "Oral Agents for the Treatment of Gestational Diabetes." *Current Diabetes Reports* 18 (11): 119. https://doi.org/10.1007/s11892-018-1093-2.

Florez, Jose C. 2017. "The Pharmacogenetics of Metformin." *Diabetologia* 60 (9): 1648–55. https://doi.org/10.1007/s00125-017-4335-y.

Foretz, Marc, Sophie Hébrard, Jocelyne Leclerc, Elham Zarrinpashneh, Maud Soty, Gilles Mithieux, Kei Sakamoto, Fabrizio Andreelli, and Benoit Viollet. 2010. "Metformin Inhibits Hepatic Gluconeogenesis in Mice Independently of the LKB1/AMPK Pathway via a Decrease in Hepatic Energy State." *Journal of Clinical Investigation* 120 (7): 2355–69. https://doi.org/10.1172/JCI40671.

Forslund, Kristoffer, Falk Hildebrand, Trine Nielsen, Gwen Falony, Emmanuelle Le Chatelier, Shinichi Sunagawa, Edi Prifti, et al. 2015. "Disentangling Type 2 Diabetes and Metformin Treatment Signatures in the Human Gut Microbiota." *Nature* 528 (7581): 262–66. https://doi.org/10.1038/nature15766.

Gandini, S., M. Puntoni, B. M. Heckman-Stoddard, B. K. Dunn, L. Ford, A. DeCensi, and E. Szabo. 2014. "Metformin and Cancer Risk and Mortality: A Systematic Review and Meta-Analysis Taking into Account Biases and Confounders." *Cancer Prevention Research* 7 (9): 867–85. https://doi.org/10.1158/1940-6207.CAPR-13-0424.

Garcia, Christine, Aaron Yao, Fabian Camacho, Rajesh Balkrishnan, and Leigh A. Cantrell. 2017. "A SEER-Medicare Analysis of the Impact of Metformin on Overall Survival in Ovarian Cancer." *Gynecologic Oncology* 146 (2): 346–50. https://doi.org/10.1016/j.ygyno.2017.05.006.

Gierach, Gretchen L, Aileen Burke, and William F Anderson. 2010. "Epidemiology of Triple Negative Breast Cancers." *Breast Disease* 32 (1–2): 5–24. https://doi.org/10.3233/BD-2010-0319.

Giovannucci, E., D. M. Harlan, M. C. Archer, R. M. Bergenstal, S. M. Gapstur, L. A. Habel, M. Pollak, J. G. Regensteiner, and D. Yee. 2010. "Diabetes and Cancer: A Consensus Report." *CA: A Cancer Journal for Clinicians* 60 (4): 207–21. https://doi.org/10.3322/caac.20078.

Gollavilli, Paradesi Naidu, Anantha Koteswararao Kanugula, Rajeswari Koyyada, Santosh Karnewar, Praveen Kumar Neeli, and Srigiridhar Kotamraju. 2015. "AMPK Inhibits MTDH Expression via GSK3β and SIRT1 Activation: Potential Role in Triple Negative Breast Cancer Cell Proliferation." *FEBS Journal* 282 (20): 3971–85. https://doi.org/10.1111/febs.13391.

Gong, Zhihong, Aaron K. Aragaki, Rowan T. Chlebowski, JoAnn E. Manson, Thomas E. Rohan, Chu Chen, Mara Z. Vitolins, et al. 2016. "Diabetes, Metformin and Incidence of and Death from Invasive Cancer in Postmenopausal Women: Results from the Women's Health Initiative." *International Journal of Cancer* 138 (8): 1915–27. https://doi.org/10.1002/ijc.29944.

Goodwin, P. J., W. R. Parulekar, K. A. Gelmon, L. E. Shepherd, J. A. Ligibel, D. L. Hershman, P. Rastogi, et al. 2015. "Effect of Metformin vs Placebo on and Metabolic Factors in NCIC CTG MA.32." *JNCI Journal of the National Cancer Institute* 107 (3): djv006–djv006. https://doi.org/10.1093/jnci/djv006.

Goodwin, Pamela J., Marguerite Ennis, Kathleen I. Pritchard, Maureen E. Trudeau, Jarley Koo, Yolanda Madarnas, Warren Hartwick, Barry Hoffman, and Nicky Hood. 2002. "Fasting Insulin and Outcome in Early-Stage Breast Cancer: Results of a Prospective Cohort Study." *Journal of Clinical Oncology* 20 (1): 42–51. https://doi.org/10.1200/JCO.2002.20.1.42.

Gormsen, L. C., E. I. Sundelin, J. B. Jensen, M. H. Vendelbo, S. Jakobsen, O. L. Munk, M. M. Hougaard Christensen, K. Brosen, J. Frokiaer, and N. Jessen. 2016. "*In Vivo* Imaging of Human 11C-Metformin in Peripheral Organs: Dosimetry, Biodistribution, and Kinetic Analyses." *Journal of Nuclear Medicine* 57 (12): 1920–26. https://doi.org/10.2967/jnumed.116.177774.

Graham, Garry G., Jeroen Punt, Manit Arora, Richard O. Day, Matthew P. Doogue, Janna K. Duong, Timothy J. Furlong, et al. 2011. "Clinical Pharmacokinetics of Metformin." *Clinical Pharmacokinetics* 50 (2): 81–98. https://doi.org/10.2165/11534750-000000000-00000.

Guimaraes, Isabella S., Nayara G. Tessarollo, Laura FRL Oliveira, Roger C. Zampier, Ian V. Silva, Cinthya Sternberg, and Leticia BA Rangel. 2015. "Abstract 2571: Metformin Inhibits Proliferation and Acts Synergistically with Paclitaxel and Doxorubicin in Triple Negative Breast Cancer Cell Lines." In *Experimental and Molecular Therapeutics*, 2571–2571. American Association for Cancer Research. https://doi.org/10.1158/1538-7445.AM2015-2571.

Gwinn, Dana M., David B. Shackelford, Daniel F. Egan, Maria M. Mihaylova, Annabelle Mery, Debbie S. Vasquez, Benjamin E. Turk, and Reuben J. Shaw. 2008. "AMPK Phosphorylation of Raptor Mediates a Metabolic Checkpoint." *Molecular Cell* 30 (2): 214–26. https://doi.org/10.1016/j.molcel.2008.03.003.

Hammond, M Elizabeth H, Daniel F Hayes, Mitch Dowsett, D Craig Allred, Karen L Hagerty, Sunil Badve, Patrick L Fitzgibbons, et al. 2010. "American Society of Clinical Oncology/College of American Pathologists Guideline Recommendations for Immunohistochemical Testing of Estrogen and Progesterone Receptors in Breast Cancer (Unabridged Version)." *Archives of Pathology & Laboratory Medicine* 134 (7): e48-72. https://doi.org/10.1043/1543-2165-134.7.e48.

Han, T., W. R. Proctor, C. L. Costales, H. Cai, R. S. Everett, and D. R. Thakker. 2015. "Four Cation-Selective Transporters Contribute to Apical Uptake and Accumulation of Metformin in Caco-2 Cell Monolayers." *Journal of Pharmacology and Experimental Therapeutics* 352 (3): 519–28. https://doi.org/10.1124/jpet.114.220350.

Han, Yamei, Zhimin Hu, Aoyuan Cui, Zhengshuai Liu, Fengguang Ma, Yaqian Xue, Yuxiao Liu, et al. 2019. "Post-Translational Regulation of Lipogenesis via AMPK-Dependent Phosphorylation of Insulin-Induced Gene." *Nature Communications* 10 (1): 623. https://doi.org/10.1038/s41467-019-08585-4.

Hara, Kenta, Kazuyoshi Yonezawa, Qing-Ping Weng, Mark T. Kozlowski, Christopher Belham, and Joseph Avruch. 1998. "Amino Acid Sufficiency and MTOR Regulate P70 S6 Kinase and EIF-4E BP1 through a Common Effector Mechanism." *Journal of Biological*

Chemistry 273 (23): 14484–94. https://doi.org/10.1074/jbc.273.23.14484.

Harbeck, Nadia, and Michael Gnant. 2017. "Breast Cancer." *The Lancet* 389 (10074): 1134–50. https://doi.org/10.1016/S0140-6736(16)31891-8.

Hardefeldt, Prue J, Senarath Edirimanne, and Guy D Eslick. 2012. "Diabetes Increases the Risk of Breast Cancer: A Meta-Analysis." *Endocrine-Related Cancer* 19 (6): 793–803. https://doi.org/10.1530/ERC-12-0242.

Hardie, D. Grahame, Fiona A. Ross, and Simon A. Hawley. 2012. "AMPK: A Nutrient and Energy Sensor That Maintains Energy Homeostasis." *Nature Reviews Molecular Cell Biology* 13 (4): 251–62. https://doi.org/10.1038/nrm3311.

Hawley, Simon A., David A. Pan, Kirsty J. Mustard, Louise Ross, Jenny Bain, Arthur M. Edelman, Bruno G. Frenguelli, and D. Grahame Hardie. 2005. "Calmodulin-Dependent Protein Kinase Kinase-β Is an Alternative Upstream Kinase for AMP-Activated Protein Kinase." *Cell Metabolism* 2 (1): 9–19. https://doi.org/10.1016/j.cmet.2005.05.009.

Hijaz, M., J. Chhina, I. Mert, M. Taylor, S. Dar, Z. Al-Wahab, R. Ali-Fehmi, T. Buekers, A. R. Munkarah, and R. Rattan. 2016. "Preclinical Evaluation of Olaparib and Metformin Combination in BRCA1 Wildtype Ovarian Cancer." *Gynecologic Oncology* 142 (2): 323–31. https://doi.org/10.1016/j.ygyno.2016.06.005.

Hirsch, H. A., D. Iliopoulos, P. N. Tsichlis, and K. Struhl. 2009. "Metformin Selectively Targets Cancer Stem Cells, and Acts Together with Chemotherapy to Block Tumor Growth and Prolong Remission." *Cancer Research* 69 (19): 7507–11. https://doi.org/10.1158/0008-5472.CAN-09-2994.

Holman, Rury R., Sanjoy K. Paul, M. Angelyn Bethel, David R. Matthews, and H. Andrew W. Neil. 2008. "10-Year Follow-up of Intensive Glucose Control in Type 2 Diabetes." *New England Journal of Medicine* 359 (15): 1577–89. https://doi.org/10.1056/NEJMoa0806470.

Hou, Yilin, Man Zhou, Jing Xie, Ping Chao, Qiyu Feng, and Jun Wu. 2017. "High Glucose Levels Promote the Proliferation of Breast Cancer Cells through GTPases." *Breast Cancer: Targets and Therapy* Volume 9 (June): 429–36. https://doi.org/10.2147/BCTT.S135665.

Howe, L. R., K. Subbaramaiah, C. A. Hudis, and A. J. Dannenberg. 2013. "Molecular Pathways: Adipose Inflammation as a Mediator of Obesity-Associated Cancer." *Clinical Cancer Research* 19 (22): 6074–83. https://doi.org/10.1158/1078-0432.CCR-12-2603.

Huang, Jia, Lin Zhang, Joel Greshock, Theresa A. Colligon, Yan Wang, Renee Ward, Dionyssios Katsaros, et al. 2011. "Frequent Genetic Abnormalities of the PI3K/AKT Pathway in Primary Ovarian Cancer Predict Patient Outcome." *Genes, Chromosomes and Cancer* 50 (8): 606–18. https://doi.org/10.1002/gcc.20883.

Hunter, Roger W., Curtis C. Hughey, Louise Lantier, Elias I. Sundelin, Mark Peggie, Elton Zeqiraj, Frank Sicheri, Niels Jessen, David H. Wasserman, and Kei Sakamoto. 2018. "Metformin Reduces Liver Glucose Production by Inhibition of Fructose-1-6-Bisphosphatase." *Nature Medicine* 24 (9): 1395–1406. https://doi.org/10.1038/s41591-018-0159-7.

Huo, J, X-H Bian, Y Huang, Z-C Miao, and L-H Song. 2017. "Inhibitory Effect and Mechanism of Metformin on Human Ovarian Cancer Cells SKOV-3 and A2780." *European Review for Medical and Pharmacological Sciences* 21 (3): 484–89. http://www.ncbi.nlm.nih.gov/pubmed/28239823.

Iliopoulos, D., H. A. Hirsch, and K. Struhl. 2011. "Metformin Decreases the Dose of Chemotherapy for Prolonging Tumor Remission in Mouse Xenografts Involving Multiple Cancer Cell Types." *Cancer Research* 71 (9): 3196–3201. https://doi.org/10.1158/0008-5472.CAN-10-3471.

Inoki, K. 2003. "Rheb GTPase Is a Direct Target of TSC2 GAP Activity and Regulates MTOR Signaling." *Genes & Development* 17 (15): 1829–34. https://doi.org/10.1101/gad.1110003.

Inoki, Ken, Michael N Corradetti, and Kun-Liang Guan. 2005. "Dysregulation of the TSC-MTOR Pathway in Human Disease." *Nature Genetics* 37 (1): 19–24. https://doi.org/10.1038/ng1494.

Jang, Soon Young, Aeree Kim, Jin Kyoung Kim, Chungyel Kim, Youl-Hee Cho, Jong-Ho Kim, Chul Hwan Kim, and Ji-Yun Lee. 2014. "Metformin Inhibits Tumor Cell Migration via Down-Regulation of MMP9 in Tamoxifen-Resistant Breast Cancer Cells." *Anticancer Research* 34 (8): 4127–34. http://www.ncbi.nlm.nih.gov/pubmed/25075039.

Jensen, Jonas B., Elias I. Sundelin, Steen Jakobsen, Lars C. Gormsen, Ole L. Munk, Jørgen Frøkiær, and Niels Jessen. 2016. "[11 C]-Labeled Metformin Distribution in the Liver and Small Intestine Using Dynamic Positron Emission Tomography in Mice Demonstrates Tissue-Specific Transporter Dependency." *Diabetes* 65 (6): 1724–30. https://doi.org/10.2337/db16-0032.

Jiralerspong, Sao, Shana L. Palla, Sharon H. Giordano, Funda Meric-Bernstam, Cornelia Liedtke, Chad M. Barnett, Limin Hsu, Mien-Chie Hung, Gabriel N. Hortobagyi, and Ana M. Gonzalez-Angulo. 2009. "Metformin and Pathologic Complete Responses to Neoadjuvant Chemotherapy in Diabetic Patients With Breast Cancer." *Journal of Clinical Oncology* 27 (20): 3297–3302. https://doi.org/10.1200/JCO.2009.19.6410.

Johnson, J. A., B. Carstensen, D. Witte, S. L. Bowker, L. Lipscombe, and A. G. Renehan. 2012. "Diabetes and Cancer (1): Evaluating the Temporal Relationship between Type 2 Diabetes and Cancer Incidence." *Diabetologia* 55 (6): 1607–18. https://doi.org/10.1007/s00125-012-2525-1.

Jung, Ji-Won, Sang-Bum Park, Soo-Jin Lee, Min-Soo Seo, James E. Trosko, and Kyung-Sun Kang. 2011. "Metformin Represses Self-Renewal of the Human Breast Carcinoma Stem Cells via Inhibition of Estrogen Receptor-Mediated OCT4 Expression." Edited by Syed A. Aziz. *PLoS ONE* 6 (11): e28068. https://doi.org/10.1371/journal.pone.0028068.

Kakar, S. 2000. "HER-2/Neu Assessment in Breast Cancer by Immunohistochemistry and Fluorescence in Situ Hybridization: Comparison of Results and Correlation with Survival." *Molecular Diagnosis* 5 (3): 199–207. https://doi.org/10.1054/modi.2000.16690.

Kalender, Adem, Anand Selvaraj, So Young Kim, Pawan Gulati, Sophie Brûlé, Benoit Viollet, Bruce E. Kemp, et al. 2010. "Metformin, Independent of AMPK, Inhibits MTORC1 in a Rag GTPase-Dependent Manner." *Cell Metabolism* 11 (5): 390–401. https://doi.org/10.1016/j.cmet.2010.03.014.

Kalinsky, Kevin, Katherine D. Crew, Susan Refice, Tong Xiao, Antai Wang, Sheldon M. Feldman, Bret Taback, et al. 2014. "Presurgical Trial of Metformin in Overweight and Obese Patients with Newly Diagnosed Breast Cancer." *Cancer Investigation* 32 (4): 150–57. https://doi.org/10.3109/07357907.2014.889706.

Kim, H.-J., B. C. Litzenburger, X. Cui, D. A. Delgado, B. C. Grabiner, X. Lin, M. T. Lewis, et al. 2007. "Constitutively Active Type I Insulin-Like Growth Factor Receptor Causes Transformation and Xenograft Growth of Immortalized Mammary Epithelial Cells and Is Accompanied by an Epithelial-to-Mesenchymal Transition Mediated by NF- B and Snail." *Molecular and Cellular Biology* 27 (8): 3165–75. https://doi.org/10.1128/MCB.01315-06.

Kim, Jinkyoung, Jiyun Lee, Soon Young Jang, Chungyeul Kim, Yoojin Choi, and Aeree Kim. 2016. "Anticancer Effect of Metformin on Estrogen Receptor-Positive and Tamoxifen-Resistant Breast Cancer Cell Lines." *Oncology Reports* 35 (5): 2553–60. https://doi.org/10.3892/or.2016.4675.

Kim, Jisun, Woosung Lim, Eun-Kyu Kim, Min-Kyoon Kim, Nam-Sun Paik, Sang-Seol Jeong, Jung-han Yoon, et al. 2014. "Phase II Randomized Trial of Neoadjuvant Metformin plus Letrozole versus Placebo plus Letrozole for Estrogen Receptor Positive Postmenopausal Breast Cancer (METEOR)." *BMC Cancer* 14 (1): 170. https://doi.org/10.1186/1471-2407-14-170.

KIM, NAM-YI, HWA-YOUNG LEE, and CHUHEE LEE. 2015. "Metformin Targets Axl and Tyro3 Receptor Tyrosine Kinases to Inhibit Cell Proliferation and Overcome Chemoresistance in Ovarian Cancer Cells." *International Journal of Oncology* 47 (1): 353–60. https://doi.org/10.3892/ijo.2015.3004.

Ko, Kwang-Pil, Seung Hyun Ma, Jae-Jeong Yang, Yunji Hwang, Choonghyun Ahn, Young-Min Cho, Dong-Young Noh, Byung-Joo Park, Wonshik Han, and Sue K. Park. 2015. "Metformin Intervention in Obese Non-Diabetic Patients with Breast Cancer: Phase II Randomized, Double-Blind, Placebo-Controlled Trial." *Breast Cancer Research and Treatment* 153 (2): 361–70. https://doi.org/10.1007/s10549-015-3519-8.

Krishan, S., D. R. Richardson, and S. Sahni. 2015. "Adenosine Monophosphate-Activated Kinase and Its Key Role in Catabolism: Structure, Regulation, Biological Activity, and Pharmacological Activation." *Molecular Pharmacology* 87 (3): 363–77. https://doi.org/10.1124/mol.114.095810.

Kumar, Sanjeev, Alexandra Meuter, Prabin Thapa, Carrie Langstraat, Shailendra Giri, Jeremy Chien, Ramandeep Rattan, William Cliby, and Viji Shridhar. 2013. "Metformin Intake Is Associated with Better Survival in Ovarian Cancer." *Cancer* 119 (3): 555–62. https://doi.org/10.1002/cncr.27706.

Landman, G. W.D., N. Kleefstra, K. J.J. van Hateren, K. H. Groenier, R. O.B. Gans, and H. J.G. Bilo. 2010. "Metformin Associated With Lower Cancer Mortality in Type 2 Diabetes: ZODIAC-16." *Diabetes Care* 33 (2): 322–26. https://doi.org/10.2337/dc09-1380.

Lashen, Hany. 2010. "Review: Role of Metformin in the Management of Polycystic Ovary Syndrome." *Therapeutic Advances in Endocrinology and Metabolism* 1 (3): 117–28. https://doi.org/10.1177/2042018810380215.

Lau, Ying-Ka Ingar, Xing Du, Vinayak Rayannavar, Benjamin Hopkins, Jacquelyn Shaw, Eliana Bessler, Tiffany Thomas, et al. 2014. "Metformin and Erlotinib Synergize to Inhibit Basal Breast Cancer." *Oncotarget* 5 (21). https://doi.org/10.18632/oncotarget.2391.

Lee, Jung-Yun, InPyo Jeon, Jae Weon Kim, Yong-Sang Song, Jae-Moon Yoon, and Sang Min Park. 2013. "Diabetes Mellitus and Ovarian Cancer Risk: A Systematic Review and Meta-Analysis of Observational Studies." *International Journal of Gynecologic Cancer* 23 (3): 402–12. https://doi.org/10.1097/IGC.0b013e31828189b2.

Lengyel, Ernst, Lacey M. Litchfield, Anirban K. Mitra, Kristin M. Nieman, Abir Mukherjee, Yilin Zhang, Alyssa Johnson, Michael Bradaric, WooSeok Lee, and Iris L. Romero. 2015. "Metformin Inhibits Ovarian Cancer Growth and Increases Sensitivity to Paclitaxel in Mouse Models." *American Journal of Obstetrics and Gynecology* 212 (4): 479.e1-479.e10. https://doi.org/10.1016/j.ajog.2014.10.026.

Lettieri Barbato, Daniele, Rolando Vegliante, Enrico Desideri, and Maria Rosa Ciriolo. 2014. "Managing Lipid Metabolism in Proliferating Cells: New Perspective for Metformin Usage in Cancer Therapy." *Biochimica et Biophysica Acta (BBA) - Reviews on Cancer* 1845 (2): 317–24. https://doi.org/10.1016/j.bbcan.2014.02.003.

Li, Ying, Jing Luo, Meng-Ting Lin, Pei Zhi, Wang-Wei Guo, Min Han, Jian You, and Jian-Qing Gao. 2019. "Co-Delivery of Metformin Enhances the Antimultidrug Resistant Tumor Effect of Doxorubicin by Improving Hypoxic Tumor Microenvironment." *Molecular Pharmaceutics* 16 (7): 2966–79. https://doi.org/10.1021/acs.molpharmaceut.9b00199.

Li, Yu, Shanqin Xu, Maria M. Mihaylova, Bin Zheng, Xiuyun Hou, Bingbing Jiang, Ogyi Park, et al. 2011. "AMPK Phosphorylates and Inhibits SREBP Activity to Attenuate Hepatic Steatosis and Atherosclerosis in Diet-Induced Insulin-Resistant Mice." *Cell Metabolism* 13 (4): 376–88. https://doi.org/10.1016/j.cmet.2011.03.009.

Libby, G., L. A. Donnelly, P. T. Donnan, D. R. Alessi, A. D. Morris, and J. M. M. Evans. 2009. "New Users of Metformin Are at Low Risk of Incident Cancer: A Cohort Study among People with Type 2 Diabetes." *Diabetes Care* 32 (9): 1620–25. https://doi.org/10.2337/dc08-2175.

Liu, Bolin, Zeying Fan, Susan M. Edgerton, Xin-Sheng Deng, Irina N. Alimova, Stuart E. Lind, and Ann D. Thor. 2009. "Metformin Induces Unique Biological and Molecular Responses in Triple Negative Breast Cancer Cells." *Cell Cycle* 8 (13): 2031–40. https://doi.org/10.4161/cc.8.13.8814.

Liu, Bolin, Zeying Fan, Susan M. Edgerton, XiaoHe Yang, Stuart E. Lind, and Ann D. Thor. 2011. “Potent Anti-Proliferative Effects of Metformin on Trastuzumab-Resistant Breast Cancer Cells via Inhibition of ErbB2/IGF-1 Receptor Interactions.” *Cell Cycle* 10 (17): 2959–66. https://doi.org/10.4161/cc.10.17.16359.

Liu, Qilin, Sheyu Li, Heng Quan, and Jianwei Li. 2014. “Vitamin B12 Status in Metformin Treated Patients: Systematic Review.” Edited by Massimo Pietropaolo. *PLoS ONE* 9 (6): e100379. https://doi.org/10.1371/journal.pone.0100379.

Liu, Yingzhao, Yan Feng, Hongmei Liu, Jianyong Wu, Yong Tang, and Qi Wang. 2018. “Real-time Assessment of Platinum Sensitivity of Primary Culture from a Patient with Ovarian Cancer with Extensive Metastasis and the Platinum Sensitivity Enhancing Effect by Metformin.” *Oncology Letters*, July. https://doi.org/10.3892/ol.2018.9223.

Lohmann, Ana Elisa, Mira F. Liebman, William Brien, Wendy R. Parulekar, Karen A. Gelmon, Lois E. Shepherd, Jennifer A. Ligibel, et al. 2017. “Effects of Metformin versus Placebo on Vitamin B12 Metabolism in Non-Diabetic Breast Cancer Patients in CCTG MA.32.” *Breast Cancer Research and Treatment* 164 (2): 371–78. https://doi.org/10.1007/s10549-017-4265-x.

Lord, Simon R., Wei-Chen Cheng, Dan Liu, Edoardo Gaude, Syed Haider, Tom Metcalf, Neel Patel, et al. 2018. “Integrated Pharmacodynamic Analysis Identifies Two Metabolic Adaption Pathways to Metformin in Breast Cancer.” *Cell Metabolism* 28 (5): 679-688.e4. https://doi.org/10.1016/j.cmet.2018.08.021.

Luft, D., R. M. Schmülling, and M. Eggstein. 1978. “Lactic Acidosis in Biguanide-Treated Diabetics.” *Diabetologia* 14 (2): 75–87. https://doi.org/10.1007/BF01263444.

Luo, Ming, Shawn G. Clouthier, Yadwinder Deol, Suling Liu, Sunitha Nagrath, Ebrahim Azizi, and Max S. Wicha. 2015. *Breast Cancer Stem Cells: Current Advances and Clinical Implications.* In, 1–49. https://doi.org/10.1007/978-1-4939-2519-3_1.

Lupia, Michela, and Ugo Cavallaro. 2017. "Ovarian Cancer Stem Cells: Still an Elusive Entity?" *Molecular Cancer* 16 (1): 64. https://doi.org/10.1186/s12943-017-0638-3.

Ma, Ji, Yan Guo, Suning Chen, Cuiping Zhong, Yan Xue, Yuan Zhang, Xiaofeng Lai, et al. 2014. "Metformin Enhances Tamoxifen-Mediated Tumor Growth Inhibition in ER-Positive Breast Carcinoma." *BMC Cancer* 14 (1): 172. https://doi.org/10.1186/1471-2407-14-172.

Mabuchi, Seiji, Hiromasa Kuroda, Ryoko Takahashi, and Tomoyuki Sasano. 2015. "The PI3K/AKT/MTOR Pathway as a Therapeutic Target in Ovarian Cancer." *Gynecologic Oncology* 137 (1): 173–79. https://doi.org/10.1016/j.ygyno.2015.02.003.

MacKenzie, Mary J., Scott Ernst, Craig Johnson, and Eric Winquist. 2012. "A Phase I Study of Temsirolimus and Metformin in Advanced Solid Tumours." *Investigational New Drugs* 30 (2): 647–52. https://doi.org/10.1007/s10637-010-9570-8.

Madiraju, Anila K., Derek M. Erion, Yasmeen Rahimi, Xian-Man Zhang, Demetrios T. Braddock, Ronald A. Albright, Brett J. Prigaro, et al. 2014. "Metformin Suppresses Gluconeogenesis by Inhibiting Mitochondrial Glycerophosphate Dehydrogenase." *Nature* 510 (7506): 542–46. https://doi.org/10.1038/nature13270.

Madiraju, Anila K., Yang Qiu, Rachel J. Perry, Yasmeen Rahimi, Xian-Man Zhang, Dongyan Zhang, João-Paulo G. Camporez, et al. 2018. "Metformin Inhibits Gluconeogenesis via a Redox-Dependent Mechanism *in Vivo*." *Nature Medicine* 24 (9): 1384–94. https://doi.org/10.1038/s41591-018-0125-4.

Mannucci, E., A. Ognibene, F. Cremasco, G. Bardini, A. Mencucci, E. Pierazzuoli, S. Ciani, G. Messeri, and C. M. Rotella. 2001. "Effect of Metformin on Glucagon-Like Peptide 1 (GLP-1) and Leptin Levels in Obese Nondiabetic Subjects." *Diabetes Care* 24 (3): 489–94. https://doi.org/10.2337/diacare.24.3.489.

Marinello, Poliana Camila, Carolina Panis, Thamara Nishida Xavier Silva, Renata Binato, Eliana Abdelhay, Juliana Alves Rodrigues, André Luiz Mencalha, et al. 2019. "Metformin Prevention of Doxorubicin Resistance in MCF-7 and MDA-MB-231 Involves Oxidative Stress

Generation and Modulation of Cell Adaptation Genes." *Scientific Reports* 9 (1): 5864. https://doi.org/10.1038/s41598-019-42357-w.

Martinez, Jessica A., Pavani Chalasani, Cynthia A. Thomson, Denise Roe, Maria Altbach, Jean-Philippe Galons, Alison Stopeck, Patricia A. Thompson, Diana Evelyn Villa-Guillen, and H-H. Sherry Chow. 2016. "Phase II Study of Metformin for Reduction of Obesity-Associated Breast Cancer Risk: A Randomized Controlled Trial Protocol." *BMC Cancer* 16 (1): 500. https://doi.org/10.1186/s12885-016-2551-3.

McCreight, Laura J., Clifford J. Bailey, and Ewan R. Pearson. 2016. "Metformin and the Gastrointestinal Tract." *Diabetologia* 59 (3): 426–35. https://doi.org/10.1007/s00125-015-3844-9.

Miller, Russell A., Qingwei Chu, Jianxin Xie, Marc Foretz, Benoit Viollet, and Morris J. Birnbaum. 2013. "Biguanides Suppress Hepatic Glucagon Signalling by Decreasing Production of Cyclic AMP." *Nature* 494 (7436): 256–60. https://doi.org/10.1038/nature11808.

Miller, Todd W, Bryan T Hennessy, Ana M González-Angulo, Emily M Fox, Gordon B Mills, Heidi Chen, Catherine Higham, Carlos García-Echeverría, Yu Shyr, and Carlos L Arteaga. 2010. "Hyperactivation of Phosphatidylinositol-3 Kinase Promotes Escape from Hormone Dependence in Estrogen Receptor-Positive Human Breast Cancer." *The Journal of Clinical Investigation* 120 (7): 2406–13. https://doi.org/10.1172/JCI41680.

Misbin, R I, L Green, B V Stadel, J L Gueriguian, A Gubbi, and G A Fleming. 1998. "Lactic Acidosis in Patients with Diabetes Treated with Metformin." *The New England Journal of Medicine* 338 (4): 265–66. https://doi.org/10.1056/NEJM199801223380415.

Momenimovahed, Zohre, Azita Tiznobaik, Safoura Taheri, and Hamid Salehiniya. 2019. "Ovarian Cancer in the World: Epidemiology and Risk Factors." *International Journal of Women's Health* Volume 11 (April): 287–99. https://doi.org/10.2147/IJWH.S197604.

Nanni, O., D. Amadori, A. De Censi, A. Rocca, A. Freschi, A. Bologna, L. Gianni, et al. 2019. "Metformin plus Chemotherapy versus Chemotherapy Alone in the First-Line Treatment of HER2-Negative Metastatic Breast Cancer. The MYME Randomized, Phase 2 Clinical

Trial." *Breast Cancer Research and Treatment* 174 (2): 433–42. https://doi.org/10.1007/s10549-018-05070-2.

Nieman, Kristin M, Hilary A Kenny, Carla V Penicka, Andras Ladanyi, Rebecca Buell-Gutbrod, Marion R Zillhardt, Iris L Romero, et al. 2011. "Adipocytes Promote Ovarian Cancer Metastasis and Provide Energy for Rapid Tumor Growth." *Nature Medicine* 17 (11): 1498–1503. https://doi.org/10.1038/nm.2492.

Owen, M R, E Doran, and A P Halestrap. 2000. "Evidence That Metformin Exerts Its Anti-Diabetic Effects through Inhibition of Complex 1 of the Mitochondrial Respiratory Chain." *The Biochemical Journal* 348 Pt 3 (June): 607–14. http://www.ncbi.nlm.nih.gov/pubmed/10839993.

Patel, Nirmesh, Daniel Weekes, Konstantinos Drosopoulos, Patrycja Gazinska, Elodie Noel, Mamun Rashid, Hasan Mirza, et al. 2018. "Integrated Genomics and Functional Validation Identifies Malignant Cell Specific Dependencies in Triple Negative Breast Cancer." *Nature Communications* 9 (1). https://doi.org/10.1038/s41467-018-03283-z.

Perou, Charles M., Therese Sørlie, Michael B. Eisen, Matt van de Rijn, Stefanie S. Jeffrey, Christian A. Rees, Jonathan R. Pollack, et al. 2000. "Molecular Portraits of Human Breast Tumours." *Nature* 406 (6797): 747–52. https://doi.org/10.1038/35021093.

Peyrat, J. P., M. M. Louchez, J. Lefebvre, J. Bonneterre, P. Vennin, A. Demaille, B. Helquet, and C. Fournier. 1993. "Plasma Insulin-like Growth Factor-1 (IGF-1) Concentrations in Human Breast Cancer." *European Journal of Cancer* 29 (4): 492–97. https://doi.org/10.1016/S0959-8049(05)80137-6.

Pollak, Michael. 2012. "The Insulin and Insulin-like Growth Factor Receptor Family in Neoplasia: An Update." *Nature Reviews Cancer* 12 (3): 159–69. https://doi.org/10.1038/nrc3215.

Qu, Chen, Weijia Zhang, Guopei Zheng, Zijuan Zhang, Jiang Yin, and Zhimin He. 2014. "Metformin Reverses Multidrug Resistance and Epithelial–Mesenchymal Transition (EMT) via Activating AMP-Activated Protein Kinase (AMPK) in Human Breast Cancer Cells." *Molecular and Cellular Biochemistry* 386 (1–2): 63–71. https://doi.org/10.1007/s11010-013-1845-x.

Queiroz, Eveline A. I. F., Stephanie Puukila, Rosangela Eichler, Sandra C. Sampaio, Heidi L. Forsyth, Simon J. Lees, Aneli M. Barbosa, Robert F. H. Dekker, Zuleica B. Fortes, and Neelam Khaper. 2014. "Metformin Induces Apoptosis and Cell Cycle Arrest Mediated by Oxidative Stress, AMPK and FOXO3a in MCF-7 Breast Cancer Cells." Edited by Masaru Katoh. *PLoS ONE* 9 (5): e98207. https://doi.org/10.1371/journal.pone.0098207.

Rattan, R., S. Giri, L. C. Hartmann, and V. Shridhar. 2011. "Metformin Attenuates Ovarian Cancer Cell Growth in an AMP-Kinase Dispensable Manner." *Journal of Cellular and Molecular Medicine* 15 (1): 166–78. https://doi.org/10.1111/j.1582-4934.2009.00954.x.

Rattan, Ramandeep, Rondell P. Graham, Jacie L. Maguire, Shailendra Giri, and Viji Shridhar. 2011. "Metformin Suppresses Ovarian Cancer Growth and Metastasis with Enhancement of Cisplatin Cytotoxicity *In Vivo*." *Neoplasia* 13 (5): 483-IN28. https://doi.org/10.1593/neo.11148.

Ricci, Francesca, Laura Brunelli, Roberta Affatato, Rosaria Chilà, Martina Verza, Stefano Indraccolo, Francesca Falcetta, et al. 2019. "Overcoming Platinum-Acquired Resistance in Ovarian Cancer Patient-Derived Xenografts." *Therapeutic Advances in Medical Oncology* 11 (January): 175883591983954. https://doi.org/10.1177/1758835919839543.

Rocha, G. Z., M. M. Dias, E. R. Ropelle, F. Osorio-Costa, F. A. Rossato, A. E. Vercesi, M. J. A. Saad, and J. B. C. Carvalheira. 2011. "Metformin Amplifies Chemotherapy-Induced AMPK Activation and Antitumoral Growth." *Clinical Cancer Research* 17 (12): 3993–4005. https://doi.org/10.1158/1078-0432.CCR-10-2243.

Ryu, Tae Young, Jiyoung Park, and Philipp E. Scherer. 2014. "Hyperglycemia as a Risk Factor for Cancer Progression." *Diabetes & Metabolism Journal* 38 (5): 330. https://doi.org/10.4093/dmj.2014.38.5.330.

Sachdev, Deepali, Julie S. Hartell, Adrian V. Lee, Xihong Zhang, and Douglas Yee. 2004. "A Dominant Negative Type I Insulin-like Growth Factor Receptor Inhibits Metastasis of Human Cancer Cells." *Journal of Biological Chemistry* 279 (6): 5017–24. https://doi.org/10.1074/jbc.M305403200.

Sachdev, Deepali, Shu-Lian Li, Julie S Hartell, Yoko Fujita-Yamaguchi, Jeffrey S Miller, and Douglas Yee. 2003. "A Chimeric Humanized Single-Chain Antibody against the Type I Insulin-like Growth Factor (IGF) Receptor Renders Breast Cancer Cells Refractory to the Mitogenic Effects of IGF-I." *Cancer Research* 63 (3): 627–35. http://www.ncbi.nlm.nih.gov/pubmed/12566306.

Sahra, I. Ben, Y. Le Marchand-Brustel, J.-F. Tanti, and F. Bost. 2010. "Metformin in Cancer Therapy: A New Perspective for an Old Antidiabetic Drug?" *Molecular Cancer Therapeutics* 9 (5): 1092–99. https://doi.org/10.1158/1535-7163.MCT-09-1186.

Sahra, I. Ben, C. Regazzetti, G. Robert, K. Laurent, Y. Le Marchand-Brustel, P. Auberger, J.-F. Tanti, S. Giorgetti-Peraldi, and F. Bost. 2011. "Metformin, Independent of AMPK, Induces MTOR Inhibition and Cell-Cycle Arrest through REDD1." *Cancer Research* 71 (13): 4366–72. https://doi.org/10.1158/0008-5472.CAN-10-1769.

Saltzman, B. S., J. A. Doherty, D. A. Hill, S. A. Beresford, L. F. Voigt, C. Chen, and N. S. Weiss. 2007. "Diabetes and Endometrial Cancer: An Evaluation of the Modifying Effects of Other Known Risk Factors." *American Journal of Epidemiology* 167 (5): 607–14. https://doi.org/10.1093/aje/kwm333.

Samarajeewa, Nirukshi U., Seungmin Ham, Fangyuan Yang, Evan R. Simpson, and Kristy A. Brown. 2011. "Promoter-Specific Effects of Metformin on Aromatase Transcript Expression." *Steroids* 76 (8): 768–71. https://doi.org/10.1016/j.steroids.2011.02.041.

Santos Guimarães, Isabella dos, Taciane Ladislau-Magescky, Nayara Gusmão Tessarollo, Diandra Zipinotti dos Santos, Etel Rodrigues Pereira Gimba, Cinthya Sternberg, Ian Victor Silva, and Leticia Batista Azevedo Rangel. 2018. "Chemosensitizing Effects of Metformin on Cisplatin- and Paclitaxel-Resistant Ovarian Cancer Cell Lines." *Pharmacological Reports* 70 (3): 409–17. https://doi.org/10.1016/j.pharep.2017.11.007.

Sayer, Robyn A., Johnathan M. Lancaster, Jennifer Pittman, Jonathon Gray, Regina Whitaker, Jeffrey R. Marks, and Andrew Berchuck. 2005. "High Insulin-like Growth Factor-2 (IGF-2) Gene Expression Is an Independent Predictor of Poor Survival for Patients with Advanced Stage Serous Epithelial Ovarian Cancer." *Gynecologic Oncology* 96 (2): 355–61. https://doi.org/10.1016/j.ygyno.2004.10.012.

Scheen, André J. 1996. "Clinical Pharmacokinetics of Metformin." *Clinical Pharmacokinetics* 30 (5): 359–71. https://doi.org/10.2165/00003088-199630050-00003.

Segal, Eric D., Amber Yasmeen, Marie-Claude Beauchamp, Joshua Rosenblatt, Michael Pollak, and Walter H. Gotlieb. 2011. "Relevance of the OCT1 Transporter to the Antineoplastic Effect of Biguanides." *Biochemical and Biophysical Research Communications* 414 (4): 694–99. https://doi.org/10.1016/j.bbrc.2011.09.134.

Sehdev, Amikar, Theodore Karrison, Yuanyuan Zha, Linda Janisch, Michelle Turcich, Ezra E. W. Cohen, Michael Maitland, et al. 2018. "A Pharmacodynamic Study of Sirolimus and Metformin in Patients with Advanced Solid Tumors." *Cancer Chemotherapy and Pharmacology* 82 (2): 309–17. https://doi.org/10.1007/s00280-018-3619-3.

Shank, Jessica J., Kun Yang, Jacob Ghannam, Lourdes Cabrera, Carolyn J. Johnston, R. Kevin Reynolds, and Ronald J. Buckanovich. 2012. "Metformin Targets Ovarian Cancer Stem Cells *in Vitro* and *in Vivo*." *Gynecologic Oncology* 127 (2): 390–97. https://doi.org/10.1016/j.ygyno.2012.07.115.

Sharma, Ankit, Shreetama Bandyopadhayaya, Kaushik Chowdhury, Tanu Sharma, Rekha Maheshwari, Amlan Das, Gopal Chakrabarti, Vipin Kumar, and Chandi C. Mandal. 2019. "Metformin Exhibited Anticancer Activity by Lowering Cellular Cholesterol Content in Breast Cancer Cells." Edited by Gokul M. Das. *PLOS ONE* 14 (1): e0209435. https://doi.org/10.1371/journal.pone.0209435.

Shi, Peiguo, Wenjing Liu, Tala, Haixia Wang, Fubing Li, Hailin Zhang, Yingying Wu, et al. 2017. "Metformin Suppresses Triple-Negative

Breast Cancer Stem Cells by Targeting KLF5 for Degradation." *Cell Discovery* 3 (1): 17010. https://doi.org/10.1038/celldisc.2017.10.

Shikata, Eriko, Rei Yamamoto, Hiroshi Takane, Chiaki Shigemasa, Tadasu Ikeda, Kenji Otsubo, and Ichiro Ieiri. 2007. "Human Organic Cation Transporter (OCT1 and OCT2) Gene Polymorphisms and Therapeutic Effects of Metformin." *Journal of Human Genetics* 52 (2): 117–22. https://doi.org/10.1007/s10038-006-0087-0.

Shu, Y, C Brown, RA Castro, RJ Shi, ET Lin, RP Owen, SA Sheardown, et al. 2008. "Effect of Genetic Variation in the Organic Cation Transporter 1, OCT1, on Metformin Pharmacokinetics." *Clinical Pharmacology & Therapeutics* 83 (2): 273–80. https://doi.org/10.1038/sj.clpt.6100275.

Song, Chang W., Hyemi Lee, Ruud P. M. Dings, Brent Williams, John Powers, Troy Dos Santos, Bo-Hwa Choi, and Heon Joo Park. 2012. "Metformin Kills and Radiosensitizes Cancer Cells and Preferentially Kills Cancer Stem Cells." *Scientific Reports* 2 (1): 362. https://doi.org/10.1038/srep00362.

Sonnenblick, Amir, Dominique Agbor-Tarh, Ian Bradbury, Serena Di Cosimo, Hatem A. Azim, Debora Fumagalli, Severine Sarp, et al. 2017. "Impact of Diabetes, Insulin, and Metformin Use on the Outcome of Patients With Human Epidermal Growth Factor Receptor 2–Positive Primary Breast Cancer: Analysis From the ALTTO Phase III Randomized Trial." *Journal of Clinical Oncology* 35 (13): 1421–29. https://doi.org/10.1200/JCO.2016.69.7722.

Soo, Jaslyn Sian-Siu, Char-Hong Ng, Si Hoey Tan, Rozita Abdul Malik, Yew-Ching Teh, Boon-Shing Tan, Gwo-Fuang Ho, et al. 2015. "Metformin Synergizes 5-Fluorouracil, Epirubicin, and Cyclophosphamide (FEC) Combination Therapy through Impairing Intracellular ATP Production and DNA Repair in Breast Cancer Stem Cells." *Apoptosis* 20 (10): 1373–87. https://doi.org/10.1007/s10495-015-1158-5.

Soranna, D., L. Scotti, A. Zambon, C. Bosetti, G. Grassi, A. Catapano, C. La Vecchia, G. Mancia, and G. Corrao. 2012. "Cancer Risk Associated with Use of Metformin and Sulfonylurea in Type 2 Diabetes: A Meta-

Analysis." *The Oncologist* 17 (6): 813–22. https://doi.org/10.1634/theoncologist.2011-0462.

Sorlie, T., C. M. Perou, R. Tibshirani, T. Aas, S. Geisler, H. Johnsen, T. Hastie, et al. 2001. "Gene Expression Patterns of Breast Carcinomas Distinguish Tumor Subclasses with Clinical Implications." *Proceedings of the National Academy of Sciences* 98 (19): 10869–74. https://doi.org/10.1073/pnas.191367098.

Tebbe, Calvin, Jasdeep Chhina, Sajad A. Dar, Kalli Sarigiannis, Shailendra Giri, Adnan R. Munkarah, and Ramandeep Rattan. 2014. "Metformin Limits the Adipocyte Tumor-Promoting Effect on Ovarian Cancer." *Oncotarget* 5 (13). https://doi.org/10.18632/oncotarget.2012.

Tessarollo, Nayara G., Isabella S. Guimaraes, Taciane Ladislau, Ian Victor Silva, and Leticia B. A. Rangel. 2015. "Abstract 2548: Metformin and Everolimus Act Synergistically with Paclitaxel against Ovarian Cancer." In *Experimental and Molecular Therapeutics*, 2548–2548. American Association for Cancer Research. https://doi.org/10.1158/1538-7445.AM2015-2548.

Thike, Aye Aye, Poh Yian Cheok, Ana Richelia Jara-Lazaro, Benita Tan, Patrick Tan, and Puay Hoon Tan. 2010. "Triple-Negative Breast Cancer: Clinicopathological Characteristics and Relationship with Basal-like Breast Cancer." *Modern Pathology* 23 (1): 123–33. https://doi.org/10.1038/modpathol.2009.145.

Torregrosa, Dolores, Pascual Bolufer, Ana Lluch, J. Antonio López, Eva Barragán, Amparo Ruiz, Vicente Guillem, Blanca Munárriz, and Javier García Conde. 1997. "Prognostic Significance of C-ErbB-2/Neu Amplification and Epidermal Growth Factor Receptor (EGFR) in Primary Breast Cancer and Theor Relation to Estradiol Receptor (ER) Status." *Clinica Chimica Acta* 262 (1–2): 99–119. https://doi.org/10.1016/S0009-8981(97)06542-X.

Tseng, Chin-Hsiao. 2015. "Metformin Reduces Ovarian Cancer Risk in Taiwanese Women with Type 2 Diabetes Mellitus." *Diabetes/Metabolism Research and Reviews* 31 (6): 619–26. https://doi.org/10.1002/dmrr.2649.

Tsilidis, K. K., J. C. Kasimis, D. S. Lopez, E. E. Ntzani, and J. P. A. Ioannidis. 2015. "Type 2 Diabetes and Cancer: Umbrella Review of Meta-Analyses of Observational Studies." *BMJ* 350 (jan02 1): g7607–g7607. https://doi.org/10.1136/bmj.g7607.

Tucker, GT, C Casey, PJ Phillips, H Connor, JD Ward, and HF Woods. 1981. "Metformin Kinetics in Healthy Subjects and in Patients with Diabetes Mellitus." *British Journal of Clinical Pharmacology* 12 (2): 235–46. https://doi.org/10.1111/j.1365-2125.1981.tb01206.x.

Urpilainen, Elina, Mikko Marttila, Ari Hautakoski, Martti Arffman, Reijo Sund, Pirjo Ilanne-Parikka, Reetta Arima, et al. 2018. "Prognosis of Ovarian Cancer in Women with Type 2 Diabetes Using Metformin and Other Forms of Antidiabetic Medication or Statins: A Retrospective Cohort Study." *BMC Cancer* 18 (1): 767. https://doi.org/10.1186/s12885-018-4676-z.

Vazquez-Martin, Alejandro, Cristina Oliveras-Ferraros, Sonia Del Barco, Begoña Martin-Castillo, and Javier A. Menendez. 2011. "The Anti-Diabetic Drug Metformin Suppresses Self-Renewal and Proliferation of Trastuzumab-Resistant Tumor-Initiating Breast Cancer Stem Cells." *Breast Cancer Research and Treatment* 126 (2): 355–64. https://doi.org/10.1007/s10549-010-0924-x.

Vigneri, Riccardo, and Laura Sciacca. 2018. *Diabetes and Cancer*. In, 1–34. https://doi.org/10.1007/978-3-319-27316-7_12-1.

Wahdan-Alaswad, Reema S., Dawn R. Cochrane, Nicole S. Spoelstra, Erin N. Howe, Susan M. Edgerton, Steven M. Anderson, Ann D. Thor, and Jennifer K. Richer. 2014. "Metformin-Induced Killing of Triple-Negative Breast Cancer Cells Is Mediated by Reduction in Fatty Acid Synthase via MiRNA-193b." *Hormones and Cancer* 5 (6): 374–89. https://doi.org/10.1007/s12672-014-0188-8.

Wang, Chih-Yuan, Shyang-Rong Shih, and Kuo-Chin Huang. 2019. "Increasing Risk of Diabetes Mellitus in Postmenopausal Women with Newly Diagnosed Primary Breast Cancer." *Journal of Diabetes Investigation*, July, jdi.13112. https://doi.org/10.1111/jdi.13112.

Wang, Shan-Bing, Kai-Jian Lei, Jia-Pei Liu, and Yu-Ming Jia. 2017. "Continuous Use of Metformin Can Improve Survival in Type 2

Diabetic Patients with Ovarian Cancer." *Medicine* 96 (29): e7605. https://doi.org/10.1097/MD.0000000000007605.

Wang, Yunshan, Junmin Wei, Li Li, Cong Fan, and Ying Sun. 2015. "Combined Use of Metformin and Everolimus Is Synergistic in the Treatment of Breast Cancer Cells." *Oncology Research Featuring Preclinical and Clinical Cancer Therapeutics* 22 (4): 193–201. https://doi.org/10.3727/096504015X14348950540999.

Wimberly, H., J. R. Brown, K. Schalper, H. Haack, M. R. Silver, C. Nixon, V. Bossuyt, L. Pusztai, D. R. Lannin, and D. L. Rimm. 2015. "PD-L1 Expression Correlates with Tumor-Infiltrating Lymphocytes and Response to Neoadjuvant Chemotherapy in Breast Cancer." *Cancer Immunology Research* 3 (4): 326–32. https://doi.org/10.1158/2326-6066.CIR-14-0133.

Woods, Angela, Stephen R Johnstone, Kristina Dickerson, Fiona C Leiper, Lee G D Fryer, Dietbert Neumann, Uwe Schlattner, Theo Wallimann, Marian Carlson, and David Carling. 2003. "LKB1 Is the Upstream Kinase in the AMP-Activated Protein Kinase Cascade." *Current Biology : CB* 13 (22): 2004–8. http://www.ncbi.nlm.nih.gov/pubmed/14614828.

Wu, Buchu, Shu Li, Lili Sheng, Jing Zhu, Liying Gu, Haoran Shen, Duanduan La, Brett D Hambly, Shisan Bao, and Wen Di. 2012. "Metformin Inhibits the Development and Metastasis of Ovarian Cancer." *Oncology Reports* 28 (3): 903–8. https://doi.org/10.3892/or.2012.1890.

Xu, Sen, Zongyuan Yang, Ping Jin, Xin Yang, Xiaoting Li, Xiao Wei, Ya Wang, et al. 2018. "Metformin Suppresses Tumor Progression by Inactivating Stromal Fibroblasts in Ovarian Cancer." *Molecular Cancer Therapeutics* 17 (6): 1291–1302. https://doi.org/10.1158/1535-7163.MCT-17-0927.

Yee, Douglas. 2018. "40 YEARS OF IGF1: Anti-Insulin-like Growth Factor Therapy in Breast Cancer." *Journal of Molecular Endocrinology* 61 (1): T61–68. https://doi.org/10.1530/JME-17-0261.

Yee, Sook Wah, Lawrence Lin, Matthew Merski, Michael J. Keiser, Aakash Gupta, Youcai Zhang, Huan-Chieh Chien, Brian K. Shoichet, and Kathleen M. Giacomini. 2015. "Prediction and Validation of Enzyme and Transporter Off-Targets for Metformin." *Journal of Pharmacokinetics and Pharmacodynamics* 42 (5): 463–75. https://doi.org/10.1007/s10928-015-9436-y.

Yin, M., J. Zhou, E. J. Gorak, and F. Quddus. 2013. "Metformin Is Associated With Survival Benefit in Cancer Patients With Concurrent Type 2 Diabetes: A Systematic Review and Meta-Analysis." *The Oncologist* 18 (12): 1248–55. https://doi.org/10.1634/theoncologist.2013-0111.

Yu, H. 2000. "Role of the Insulin-Like Growth Factor Family in Cancer Development and Progression." *Journal of the National Cancer Institute* 92 (18): 1472–89. https://doi.org/10.1093/jnci/92.18.1472.

Zakikhani, Mahvash, Marie-José Blouin, Esther Piura, and Michael N. Pollak. 2010. "Metformin and Rapamycin Have Distinct Effects on the AKT Pathway and Proliferation in Breast Cancer Cells." *Breast Cancer Research and Treatment* 123 (1): 271–79. https://doi.org/10.1007/s10549-010-0763-9.

Zendehdel, K. 2003. "Cancer Incidence in Patients With Type 1 Diabetes Mellitus: A Population-Based Cohort Study in Sweden." *CancerSpectrum Knowledge Environment* 95 (23): 1797–1800. https://doi.org/10.1093/jnci/djg105.

Zhang, Rongrong, Ping Zhang, Hong Wang, Dongming Hou, Wentao Li, Guishan Xiao, and Chenwei Li. 2015. "Inhibitory Effects of Metformin at Low Concentration on Epithelial–Mesenchymal Transition of CD44+CD117+ Ovarian Cancer Stem Cells." *Stem Cell Research & Therapy* 6 (1): 262. https://doi.org/10.1186/s13287-015-0249-0.

Zhang, Z.-J., and S. Li. 2014. "The Prognostic Value of Metformin for Cancer Patients with Concurrent Diabetes: A Systematic Review and Meta-Analysis." *Diabetes, Obesity and Metabolism* 16 (8): 707–10. https://doi.org/10.1111/dom.12267.

Zhao, Xiao-Bo, and Guo-Sheng Ren. 2016. "Diabetes Mellitus and Prognosis in Women with Breast Cancer." *Medicine* 95 (49): e5602. https://doi.org/10.1097/MD.0000000000005602.

Zhao, Yannan, Chengcheng Gong, Zhonghua Wang, Jian Zhang, Leiping Wang, Sheng Zhang, Jun Cao, et al. 2017. "A Randomized Phase II Study of Aromatase Inhibitors plus Metformin in Pre-Treated Postmenopausal Patients with Hormone Receptor Positive Metastatic Breast Cancer." *Oncotarget* 8 (48). https://doi.org/10.18632/oncotarget.20478.

Zheng, Ya, Jie Zhu, Haiyan Zhang, Yanmei Liu, and Hong Sun. 2018. "Metformin Inhibits Ovarian Cancer Growth and Migration *in Vitro* and *in Vivo* by Enhancing Cisplatin Cytotoxicity." *American Journal of Translational Research* 10 (10): 3086–98.

Zordoky, Beshay N. M., Diana Bark, Carrie L. Soltys, Miranda M. Sung, and Jason R. B. Dyck. 2014. "The Anti-Proliferative Effect of Metformin in Triple-Negative MDA-MB-231 Breast Cancer Cells Is Highly Dependent on Glucose Concentration: Implications for Cancer Therapy and Prevention." *Biochimica et Biophysica Acta (BBA) - General Subjects* 1840 (6): 1943–57. https://doi.org/10.1016/j.bbagen.2014.01.023.

INDEX

A

D

H

I

J

K

L

M

N

O

P

R

S

U

V

W